I0042134

Organic and Inorganic Light Emitting Diodes

This book covers a comprehensive range of topics on the physical mechanisms of LEDs (light emitting diodes), scattering effects, challenges in fabrication and efficient enhancement techniques in organic and inorganic LEDs. It deals with various reliability issues in organic/inorganic LEDs like trapping and scattering effects, packaging failures, efficiency droops, irradiation effects, thermal degradation mechanisms, and thermal degradation processes.

Features:

- Provides insights into the improvement of performance and reliability of LEDs.
- Highlights the optical power improvement mechanisms in LEDs.
- Covers the challenges in fabrication and packaging of LEDs.
- Discusses pertinent failures and degradation mechanisms.
- Includes droop minimization techniques.

This book is aimed at researchers and graduate students in LEDs, illumination engineering, optoelectronics, and polymer/organic materials.

Organic and Inorganic Light Emitting Diodes

Reliability Issues and Performance Enhancement

Edited by
T. D. Subash, J. Ajayan and Wladek Grabinski

CRC Press
Taylor & Francis Group
Boca Raton London New York

CRC Press is an imprint of the
Taylor & Francis Group, an **informa** business

Cover image: © Shutterstock

First edition published 2023
by CRC Press
6000 Broken Sound Parkway NW, Suite 300, Boca Raton, FL 33487-2742

and by CRC Press
4 Park Square, Milton Park, Abingdon, Oxon, OX14 4RN

CRC Press is an imprint of Taylor & Francis Group, LLC

© 2023 selection and editorial matter, T. D. Subash, J. Ajayan and Wladek Grabinski; individual chapters, the contributors

Reasonable efforts have been made to publish reliable data and information, but the author and publisher cannot assume responsibility for the validity of all materials or the consequences of their use. The authors and publishers have attempted to trace the copyright holders of all material reproduced in this publication and apologize to copyright holders if permission to publish in this form has not been obtained. If any copyright material has not been acknowledged please write and let us know so we may rectify in any future reprint.

Except as permitted under U.S. Copyright Law, no part of this book may be reprinted, reproduced, transmitted, or utilized in any form by any electronic, mechanical, or other means, now known or hereafter invented, including photocopying, microfilming, and recording, or in any information storage or retrieval system, without written permission from the publishers.

For permission to photocopy or use material electronically from this work, access www.copyright.com or contact the Copyright Clearance Center, Inc. (CCC), 222 Rosewood Drive, Danvers, MA 01923, 978-750-8400. For works that are not available on CCC please contact mpkbookspermissions@tandf.co.uk

Trademark notice: Product or corporate names may be trademarks or registered trademarks and are used only for identification and explanation without intent to infringe.

ISBN: 9781032375175 (hbk)
ISBN: 9781032375182 (pbk)
ISBN: 9781003340577 (ebk)

DOI: 10.1201/9781003340577

Typeset in Times
by Newgen Publishing UK

Contents

About the Editors

T. D. Subash is an academic veteran, technocrat-cum-avid researcher, mentor, innovation and entrepreneurship cell. He is currently a full-time professor, Department of Electronics and Communication Engineering, and Dean, Research & Development, VISAT Engineering College, Ernakulam, Kerala, India. He is also serving as the Global Strategy Representative for India, IEEE Photonics Society, USA. He was an active senior member of IEEE and Founding Chairman of IEEE Photonics Society Madras Chapter till 2020. He completed his Bachelor of Engineering in Electronics and Communication Engineering and Masters of Engineering in Embedded System Technologies from Anna University, India, in 2008 and 2011, respectively. He completed his PhD in Nanoelectronics from Anna University, Chennai, in 2016. He enjoys teaching and research. He has 56 publications in international and national journals and 28 papers in international and national conferences in the area of Nanoelectronics, Nanoscale Device Modelling, Nanotechnology and Wireless Sensor Networks. He also filed four patents to his credit. He is the recognized research supervisor of Anna University, Chennai, and APJ Abdul Kalam Technological University, Kerala. He serves as the active member of editorial board/reviewer board of various international journals.

J. Ajayan received his B.Tech degree in Electronics and Communication Engineering from Kerala University in 2009, M.Tech and PhD degrees in Electronics and Communication Engineering from Karunya University, Coimbatore, India, in 2012 and 2017, respectively. He is an associate professor in the Department of Electronics and Communication Engineering at SR University, Telangana, India. He has published more than 100 research articles in various journals and international conferences. He has published one book, more than ten book chapters, and two patents. He is a reviewer of more than 30 journals. He is a Guest Editor–Special Issue on P2P Computing for Beyond 5G Network (B5G) and Internet-of-Everything (IoE) by *Peer-to-Peer Networking and Applications*, Springer, and Special Issue on Energy Harvesting Devices, Circuits and Systems for Internet of Things by *Microelectronics Journal*. He has served as a member of the technical advisory/reviewer committee on more than ten conferences. His areas of interest are microelectronics, semiconductor devices, nanotechnology, RF integrated circuits and photovoltaics.

Wladek Grabinski received his PhD degree from the Institute of Electron Technology, Warsaw, Poland, in 1991. From 1991 to 1998 he was a research assistant at the Integrated Systems Lab, ETH Zürich, Switzerland, supporting the CMOS and BiCMOS technology developments by electrical characterization of the processes and devices. From 1999 to 2000, he was with LEG, EPF Lausanne, and was engaged in the compact MOSFET model developments supporting numerical device simulation and parameter extraction. Later, he was a technical staff engineer at Motorola, and subsequently at Freescale Semiconductor, Geneva Modeling Center, Switzerland. He is now a consultant responsible for modeling, characterization and parameter extraction of MOS transistors for the design of RF CMOS circuits. He is currently consulting on the development of next-generation compact models for the nanoscaled technology very large scale integration (VLSI) circuit simulation. His current research interests are in high-frequency characterization, compact modeling and its Verilog-A standardization as well as device numerical simulations of MOSFETs for analog/RF low-power IC applications. He is an editor of the reference modeling book *Transistor Level Modeling for Analog/RF IC Design* and also authored or coauthored more than 50 papers. Wladek is a member of ESSDERC TPC Track4: "Device and circuit compact modeling" as well as serving as a member of the IEEE EDS Compact Modeling Technical Committee, organization committee of ESSDERC/ESSDERC, TPC of SBMicro, SISPAD, MIXDES Conferences; reviewer of the IEEE TED, IEEE MWCL, IJNM, MEE, MEJ. He also continued as European representative to the ITRS Modeling and Simulation working group. He was a Member At Large of Swiss IEEE ExCom and also mentored the EPFL IEEE Student Branch acting as its Interim Branch Counselor. Wladek is involved in activities of the MOS-AK Association and serves as a coordinating manager since 1999.

Contributors

J. Ajayan, SR University, Warangal, India

Seemesh Bhaskar, STAR Laboratory, Sri Sathya Sai Institute of Higher Learning, Prasanthi Nilayam, Puttaparthi, Anantapur, Andhra Pradesh, India

Sagar Bhattarai, Arunachal University of Studies, Knowledge City, Namsai, Arunachal Pradesh, India

Shalu C., School of Engineering & Technology, IFTM University, Moradabad, India

G. Dhivyasri, Dr. N. G. P Institute of Technology, Coimbatore, India

N. Hemalatha, Institute of Electronics and Communication Engineering, Saveetha School of Engineering (SIMATS), Chennai, India

Binola K. Jebalin I. V., Karunya Institute of Technology and Sciences, Coimbatore, India

Ajay Kumar Mahato, National Institute of Technology, Raipur, India

Nesa Majidzadeh, University of Tabriz, Tabriz, Iran

Sankararao Majji, GRIET, Hyderabad, India

M. Manikandan, Presidency University, Bangalore, India

Hossein Movla, University of Tabriz, Tabriz, Iran

Deboraj Muchahary, National Institute of Technology, Raipur, India

D. Nirmal, Karunya Institute of Technology and Sciences, Coimbatore, India

Tulasi Radhika Patnala, Shanax Technologies, Hyderabad, Telangana, India

Moorthi Pichumani, Sri Ramakrishna Engineering College, Tamilnadu, India.

Joseph Anthony Prathap, Presidency University, Bangalore, India

M. Sundar Rajan, Institute of Technology, Arbaminch University, Ethiopia

Sai Sathish Ramamurthy, STAR Laboratory, Sri Sathya Sai Institute of Higher Learning, Prasanthi Nilayam, Puttaparthi, Anantapur, Andhra Pradesh, India

B. A. Saravanan, Sri Ramakrishna Engineering College, Tamilnadu, India.

Arvind Sharma, National Institute of Technology, Arunachal Pradesh, India

Shakti Sindhu, Oorja Technical Services Pvt. Ltd, Indore, India

T. D. Subash, VISAT Engineering College, Kerala, India

Vinodhini Subramaniyam, Sri Ramakrishna Engineering College, Coimbatore, Tamilnadu, India.

Preface

There has been a long-standing interest within the development of solid-state light-emitting devices as they need proved to be more efficient than the traditional tungsten filament light bulbs. For more than three decades, they have been utilized in various areas of applications like communication systems and lightning applications due to their longer life, higher efficiency, environmental sustainability and lower cost. These applications have placed stringent demands on improvements within the performance of LEDs. Specific expertise like understanding the materials, structure and composition thoroughly is required so as to get precise and reproducible results. In summary, LEDs have gone from infancy to adolescence leading to a continuing pressure on manufacturers to supply higher performance devices during a shorter time. The necessity for efficient light emitting diodes drives research into new direct-gap semiconductors to act as the active materials. Physical objects have now become an ecosystem of information shared between devices that are wearable, portable, and even implantable, making our lives technology and data enriched. Virtually in every industry, the machines have also gone to a subsequent level. Now it's more like they need their own brains, in terms of one just has got to let the machine know what to be done and the way and it will be at the user's beck and call. This has enabled the industry in reaching its desired production goals smoothly. Smart lighting solutions have touched numerous aspects of lives, nearly leaving no stone unturned. Healthcare services have improved with major degrees over the years with better patient care standards, but little has been done to explore the utilization of smart lights for healthcare. This book covers a comprehensive range of topics on the physical mechanisms of LED, scattering effects, challenges in fabrication and efficient enhancement techniques in organic and inorganic LEDs. Nowadays, due to the due to the rapid growth of design and manufacturing technology, there are as many consumer products with high reliability and high performance. This book deals with various reliability issues in organic/inorganic LEDs like trapping and scattering effects, packaging failures, efficiency droops, irradiation effects, thermal degradation mechanisms, etc. This book also provides insights into the improvement of performance and reliability of LEDs.

1 Fundamental Physics of Light Emitting Diodes
Organic and Inorganic Technology

Deboraj Muchahary, Sagar Bhattarai, Arvind Sharma and Ajay Kumar Mahato

CONTENTS

DOI: 10.1201/9781003340577-1

1.1   INTRODUCTION TO LEDS

Light is emitted from solid-state material while exposed to an electric power source. The phenomenon is called electroluminescence. There is a very virtuous advantage of this property in comparison to incandescence as the later one is emitted by a material heated to high temperatures, typically >750°C. Henry Joseph Round first discovered that light can be emitted from silicon carbide (SiC) crystal and the light emitting diode (LED) was born [1]. The emission of light was observed from a Schottky contact between SiC and metal, which have a rectifying current–voltage characteristics. The emission of light from inside the semiconductor material is mainly due to the recombination of charge carriers, i.e., electrons and holes. The recombination process is mainly two types–radiative and non-radiative recombination. In general, the radiative type generates light due to the recombination process, which is the major phenomenon governing the LED. On the other hand, the non-radiative type of recombination involves the generation of no light and cannot be completely removed from the material as well. The radiative type is more preferred for LED operation.

The light emission range of LED can be broadly classified into two categories: visible and ultraviolet (UV). The UV light can be further classified into three broad ranges [2,3]:

•   The wavelength range from 400 nm to 320 nm is called the UVA range.
•   The wavelength range from 320 nm to 280 nm is called the UVB range.
•   The wavelength range from 280 nm to 100 nm is called the UVC range.

UV light is used for different applications in real life. For example, UVA and UVB are used to control a chemical reaction, UVC is used to make photochemical changes to the nucleic acids of microorganism etc. Moreover, photoionization, water purification, three-dimensional (3D) printing, and optical lithography are major areas where a UV light source is used. Over the past few decades, several different inorganic materials, such as *zinc oxide (ZnO)*, *aluminiumgallium nitride (AlGaN)*, *indium gallium nitride (InGaN)*, *ZnSe* etc., have been widely explored for the advancement of UV and visible LED applications [2, 4–8]. This technology has overcome several limitations of the conventional mercury-lamp-based light source. In addition, group-III nitride-based LED technology uses eco-friendly materials.

1.2 SEMICONDUCTOR FUNDAMENTALS

1.2.1 ENERGY BANDS

The crystal structure of a semiconductor consists of several atoms of that material. Discussion of atoms without energy levels of the orbitals and momentum of the charge carriers is meaningless. In a semiconductor, energy and momentum of charge carriers are related and is crucial for the exclusive study of electrical and optical mechanisms. Material upon interaction with phonon and photon showcases conservation of momentum and energy of electrons and holes. Moreover, important parameters such as group velocity and effective mass of electrons and holes can be calculated based on this relationship. A band of energy in a material is a bunch of discrete energy levels sheltering electrons and holes. The energy–momentum $(E_n$–$p)$ relationship is described by the popular Schrodinger equation as given below.

$$H\psi = E_n \psi \tag{1.1}$$

In the equation Ψ, H and E_n stand for the wavefunction, Hamiltonian and energy, respectively. H represents the total energy operators, which consists of kinetic (KE) and potential energies (PE) of the carrier.

$$H = \frac{p_m^2}{2m_c^*} + V_p \tag{1.2}$$

The energy band effective mass of electron, momentum and PE are denoted by m_c^*, p_m and V_p, r respectively. The term h is for Plank's constant. The momentum can be replaced by its operator

$$p_m = -\left(\frac{ih\nabla}{2\pi}\right) \tag{1.3}$$

Replacing the expressions in equation 1.1 and equation 1.2, the Schrodinger equation can be obtained.

$$\left[-\left(\frac{h\nabla}{4m_c^*\pi}\right)^2 + V_p\right]\psi = E_n \psi \tag{1.4}$$

For an isolated atom of a material, if the V_p of an electron circulating in an orbit is provided in equation 4, the corresponding energy eigen value can be obtained. This justifies the fact that the isolated atom of a material consists of several discrete energy states. For example, the ground energy state of an individual hydrogen atom has an energy eigen value of 13.6 eV. However, when two atoms come into proximity, the electric field effect of one atom falls on the other atom. Due to this effect, each discrete energy state in an atom starts splitting into several smaller energy states. The split energy states constitute a band of energy called the allowed energy states. However, in each semiconductor there is a region between two energy bands where allowed states of energy are not present known as the forbidden region or energy band gap (E_g), which is essential for analyzing semiconductor devices. The lower and upper energy bands of a forbidden region are known as the valence (VB) and conduction band (CB), respectively. In addition, E_c and E_v denote energy at the bottom edge of CB and top edge VB, respectively.

1.2.2 Types of Semiconductors

Naturally available materials can be classified into insulators, semiconductors and conductors based on their electrical conductivity property. In semiconductors, the energy band gap is small compared to insulators and free charge carriers constitute electric current conduction in such materials. Based on momentum alignment of charge particles between the CB minima and the VB maxima, semiconductors are classified as direct and indirect band-gap types, which are shown in Figure 1.1(A)–(B). In a direct band-gap semiconductor, both maxima and minima of VB and CB, respectively, lie at a particular point on the momentum axis. On the other hand, in the indirect semiconductor, maxima and minima lie at different positions of the momentum axis.

Examples of the former type include GaAs, InAs, ZnO, GaN etc. while Si, Ge, and SiC etc. are semiconductors of the later type. To obtain better control on the conductivity of semiconducting materials, doping of other materials is adopted. A semiconductor doped with materials of the other type possesses additional free electrons in the CB or free holes in the VB. The additional free carriers in the CB are enhanced by ionization of imperfect donor atoms whereas those in the VB are generated from ionization of acceptor atoms doped in the material. This type of semiconductor is called an extrinsic semiconductor. Extrinsic semiconductors may be n-type or p-type depending upon the type of impurity doped in it.

1.2.3 Approximation of Charge Carrier Concentration

The charge carrier concentration symbolized by n for electrons (p for holes), is the number of free electrons (holes) that occupies unit volume of the CB (VB). The electron concentration inside a semiconducting material is the integration of the product of the density of energy states of CB($N(E_n)$) and the probability of its occupancy $F(E_n)$ over the entire CB as illustrated in Figure 1.2. The probability of occupancy of energy levels in a semiconductor is defined by Fermi–Dirac distribution function, which strongly depends on energy and temperature [9].

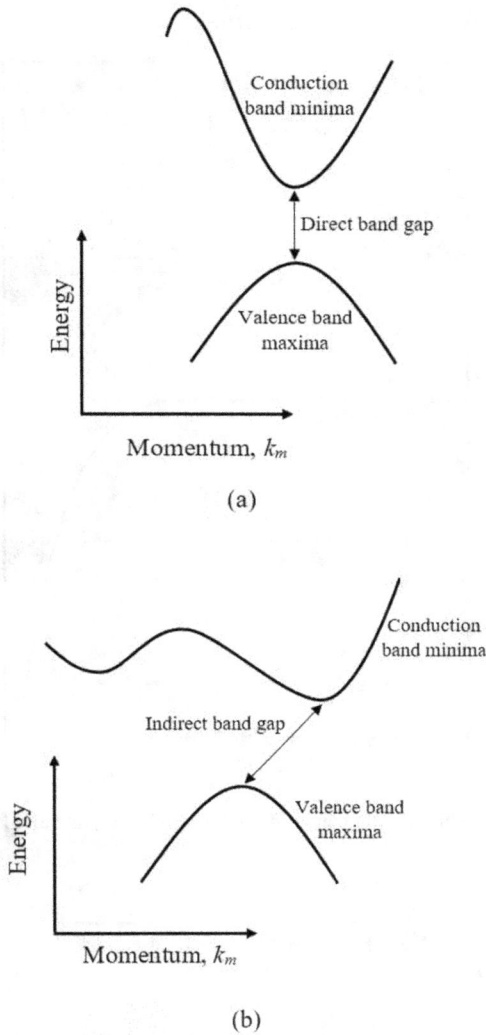

FIGURE 1.1 Types of energy band gap in semiconductor (a) direct type and (b) indirect type.

$$F(E_n) = \frac{1}{\left(1 + exp\left(\dfrac{E_n - E_{nF}}{kT}\right)\right)} \qquad (1.5)$$

where k and T denote Boltzmann's constant and temperature, respectively. The density of energy states used for the calculation of carrier density is assumed to be located at a region near the bottom edge of the CB and is given below [1].

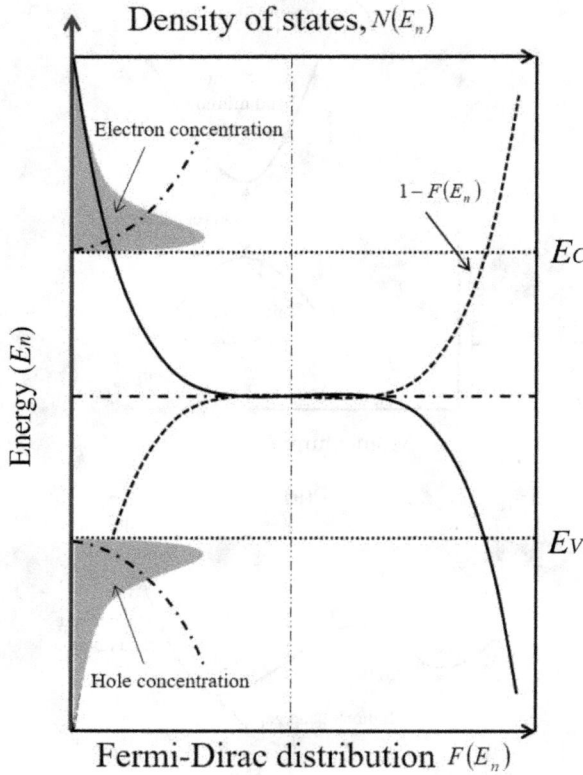

FIGURE 1.2 Relationship among carrier density, density of state and Fermi Dirac function.

$$N\left(E_n\right) = B_c 16\pi \left(\frac{\sqrt{m_c^*}}{h}\right)^3 \sqrt{\left(E_n - E_c\right)} \tag{1.6}$$

In the equation, B_c stands for number of equivalent minima in the CB. As $N(E_n)\alpha(E_n - E_c)^{0.5}$ the density is least (zero) at the bottom edge of the CB. From Figure 1.2, the probability of presence is 50% for both electrons and holes at the energy level (E_{nF}). This energy level is known as the Fermi energy level. This indicates the Fermi energy level can be evaluated based on the charge neutrality condition, i.e., the net charge is zero at that energy level. Similarly, for the calculation of hole concentration in the material, the density of state near the top edge of the VB and the probability of its occupancy by hole, i.e., $1-F(E_n)$ is multiplied and integrated over the entire VB. At this point we are familiar with the method of calculating the charge carrier concentration using the Fermi integral. However, the same can be obtained using Boltzmann's approximation of the charge carrier concentration for low and moderately doped semiconductors, which are known as non-degenerated. In such a semiconductor E_{nF} lies within the forbidden region and the separation E_{nF}-E_c is large (several times the

kT). The Boltzmann approximation provides an equation for electron concentration as given below.

$$n = N_c exp\left[\frac{-\left(E_c - E_{nF}\right)}{kT}\right] \tag{1.7}$$

Similarly, the hole concentration can be written as given below.

$$p = N_v exp\left[\frac{-\left(E_{nF} - E_v\right)}{kT}\right] \tag{1.8}$$

The addition of donor (acceptor) dopants in an intrinsic semiconductor accumulates an extra number of electrons (holes) to the intrinsic concentration. The dopants are ionized and contribute extra charge (positive if donor type and negative if acceptor type dopants) in the semiconductor. In general, the intrinsic electron concentration in an n-type semiconductor is very small and thus negligible. Therefore, the net concentration of the electron is approximately equal to the ionized donor impurity concentration (N_{DON}) according to the charge neutrality condition.

$$n \approx N_{DON} \tag{1.9}$$

Similarly, intrinsic hole concentration in a p-type semiconductor is negligible. The net hole concentration is thus approximately equal to ionized acceptor impurity concentration (N_{ACCP}).

$$p \approx N_{ACCP} \tag{1.10}$$

Using equations 1.7–1.10 the Fermi energy level with reference to the CB minima and the VB maxima for n-type and p-type semiconductors, respectively, can be evaluated as given below:

$$E_{nF} = E_c - kT \ ln\left(\frac{N_c}{n}\right); \quad \text{for } n\text{-type} \tag{1.11}$$

$$E_{pF} = E_v + kT \ ln\left(\frac{N_v}{p}\right); \quad \text{for } p\text{-type} \tag{1.12}$$

The conduction band effective density of state (DOS), N_c/valence band effective DOS, N_v is dependent on effective DOS mass for CB ($m_c^* \ m$)/effective DOS mass for VB ($m_v^* \ m$) and the number of equivalent minima in CB ($M_{eqc} \ M$)/number of equivalent maxima in VB (M_{eqv}). This dependency relationship is shown in the equations below.

$$N_c = 2M_{eqc} \left(\frac{2\pi m_c^* kT}{h^2} \right)^{\frac{3}{2}} \tag{1.13}$$

$$N_v = 2M_{eqv} \left(\frac{2\pi m_v^* kT}{h^2} \right)^{\frac{3}{2}} \tag{1.14}$$

For an intrinsic semiconductor, the position of the Fermi level (E_{Fin}) is at the vicinity of the mid-gap of the energy band. In these types of semiconductors, Fermi energy levels represented by equations (1.11) and (1.12) are equal, which gives rise to the following expression.

$$E_{Fin} = \left(\frac{Ec + Ev}{2} \right) + \frac{kT}{2} ln \left(\frac{Nv}{Nc} \right) \tag{1.15}$$

Fermi energy levels for intrinsic, n-type and p-type semiconductors are depicted in Figure 1.3.

In the case of an intrinsic semiconductor under steady state condition, the bounded electron in the VB continuously gets excited to CB by thermal agitation leaving behind the same number of holes. Some of these electrons recombine with the holes in the VB in a very short interval of time resulting in a net intrinsic concentration of charge carriers (n_i).

$$n_i = \sqrt{N_C N_V}\ exp \left(-\frac{E_g}{2kT} \right) \tag{1.16}$$

As $F(E_n)$ depends on the Fermi energy level, which is different for intrinsic, n-type and p-type semiconductors the values of n_i, n and p are different. However, in the case of a non-degenerate semiconductor, the law of mass action applies, i.e., $n_i^2 = np$ while for degenerate type $pn \neq n_i^2$ under thermal equilibrium.

1.2.4 ELECTROSTATIC POTENTIAL

In previous sections, the approximation of the electron and hole concentration is discussed. The electrostatic potential and the electric field distribution inside the semiconductor is pivotal for analysis of different characteristics of several devices. Poisson's equation shown below is a platform to analyze the electrostatic potential profile inside the material.

$$\frac{d^2V}{dx^2} = -\frac{\rho}{\varepsilon} \tag{1.17}$$

The dielectric constant of the material used, potential across and charge density inside the region of interest of a semiconductor are denoted by ε, V, and ρ, respectively. Net

(a)

(b)

(c)

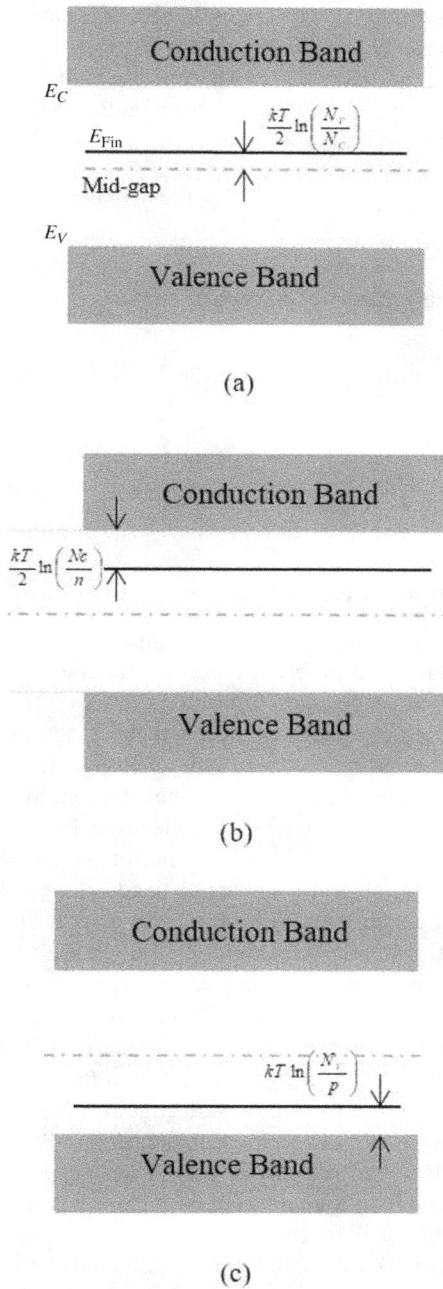

FIGURE 1.3 Position of Fermi Energy level in different types of non-degenerated semiconductors: (a) Intrinsic; (b) n-type and (c) p-type.

charge density ρ is the summation of charges on n, p, N_{DON} and N_{ACCP}. Equation (1.17) is one dimensional only, but it can be extended to higher dimension. Derivation of the equation originates from the Gauss law or Maxwell's equation. The law states that net electric flux emerging from a closed surface is equal to the total charge enclosed by the surface. Maxwell's equation corresponding to the Gauss law in three dimensions is given below.

$$\nabla.D = \rho \qquad (1.18)$$

The symbol D represents electric displacement and is given as below.

$$D = \varepsilon E = -\varepsilon \frac{dV}{dx} \qquad (1.19)$$

This equation can be used to analyze electric field distribution inside the space charge region of a p-n junction. However, to solve the Poisson equation, a proper boundary condition is required.

1.2.5 SEMICONDUCTOR P-N HOMOJUNCTION

The junction between an n-type and p-type semiconductor is achieved using a diffused method involving two different types of semiconductors. In such an architecture, alignment of their respective Fermi levels takes place. Accordingly, the CB and VB of the two types are positioned around the junction. In a homojunction, both sides of the junction are made of the same material. The electrons from n-side diffuses to p-side to maintain its equilibrium concentration. Similarly, majority holes diffuse but in the opposite direction. Diffused electrons however leave behind positively charged immobile donor ions at the vicinity of the junction in n-side. On the other side of the junction, negatively charged fixed ions get accumulated thereby establishing an internal electric field from n-type to p-type. Further, diffusion of the carrier is restricted by this internal electric field. Thus, a region is established around the junction termed as the depletion region, where free charged carriers are drifted by the build electric field. The CB minima and VB maxima energy levels vary according to the potential profile generated inside the depletion region. The depletion region edge in the n-side of the junction is at higher potential than at the edge in the p-side. As a result, E_c and E_v increase from the n-side edge to the p-side edge of the space charge region. Formation of the p-n junction is depicted in Figure 1.4.

Assuming the interface as the point of origin, the edge of the depletion region spread towards p-side and n-side are denoted by W_p and W_n, respectively. The electric potential being maximum at W_n decreases gradually inside the depletion region and becomes null at the edge $(-W_p)$. Accordingly, the potential energy given by $-qV$, bands down at the p-side edge and bands up at the n-side edge of the depletion region. Under thermal equilibrium, the net current is null, which justifies that both the Fermi energy level of n-type and p-type are aligned and stable throughout the device.

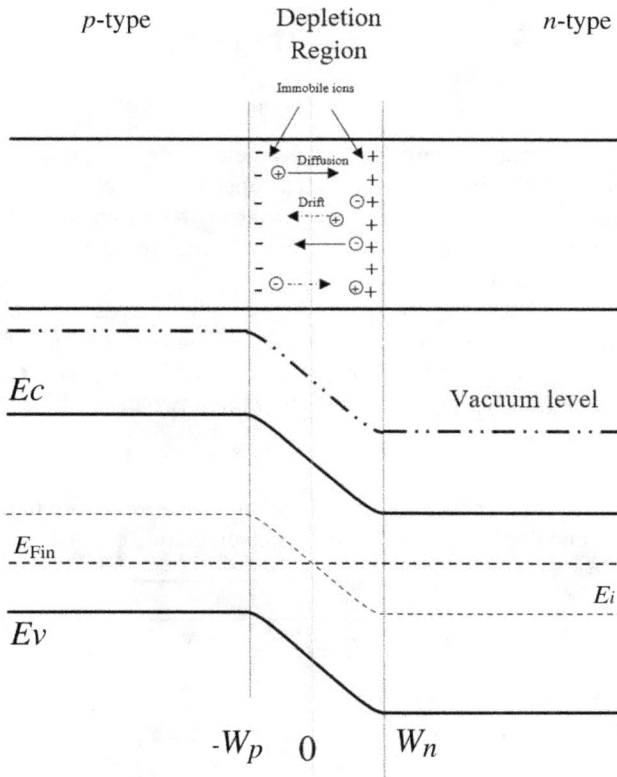

FIGURE 1.4 Formation of a semiconductor *p-n* junction (solid arrowhead line shows diffusion and dotted arrowhead line shows drift of carriers).

1.2.6 Characterizing Parameters of the Homojunction

Some major *p-n* junction characterizing parameters are evaluated from Figure 1.4 above.

(a) **Built-in potential (V_{bi}):** It is the summation of potential drop in *n*-side (V_{nbi}) and *p*-side (V_{pbi}) of the junction. The potential can be written in terms of electron (n_{te}) and hole (p_{te}) concentrations under thermal equilibrium.

$$V_{nbi} = \frac{kT}{q} \ln\left(\frac{n_{te}}{n_i}\right) \qquad (1.20)$$

$$V_{pbi} = \frac{kT}{q} \ln\left(\frac{p_{te}}{n_i}\right) \qquad (1.21)$$

$$V_{bi} = V_{nbi} + V_{pbi} = \frac{kT}{q} \ln\left(\frac{n_{te}P_{te}}{n_i^2}\right) \qquad (1.22)$$

Built-in potential is an essential parameter to describe operation in various devices especially in the solar cell. The open circuit voltage of a *p-n* junction solar cell device is directly proportional to the parameter. The V_{bi} can be determined by capacitance measurement with respect to applied biasing (description is below).

(b) **Depletion width** (*W*)**:** The space charge region across the *p-n* junction has two parts spread over a range of width on the material. One part has a width of W_n towards *n*-side whereas the other has width of W_p in *p*-side of the junction. The depletion width is the summation of these two parts of the region.

$$W = W_n + W_p \qquad (1.23)$$

In the equation partial depletion widths are expressed in terms of built-in potential and doping concentrations as given below.

$$W_n = \sqrt{\frac{2\varepsilon V_{bi}}{q}\frac{N_{DON}}{N_{ACCP}\left(N_{ACCP}+N_{DON}\right)}} \qquad (1.24)$$

$$W_p = \sqrt{\frac{2\varepsilon V_{bi}}{q}\frac{N_{ACCP}}{N_{DON}\left(N_{ACCP}+N_{DON}\right)}} \qquad (1.25)$$

(c) **Electric field inside the depletion region** (*E*)**:** The electric field has its maximum magnitude (E_{MAX}) at the junction or point of origin and decreases linearly on either side of it. The electric field at the edges of depletion region and inside the bulk of the material are very small and thus it can be assumed null. Distribution of *E* across the junction is shown in Figure 1.5. The magnitude of maximum electric field for a *p-n* homojunction can be represented as given below.

$$\left|E_{MAX}\right| = \frac{qN_{DON}W_p}{\varepsilon}; \quad \text{at } x = 0 \qquad (1.26)$$

(d) **Junction capacitance:** The capacitance per unit area of the junction (one sided abrupt) denoted by C_{depl} is due to the electric field in the space charge region. Two oppositely charged layers at the edges of the space charge region constitutes a parallel plate-like capacitor. The capacitance is temperature and applied potential dependent as presented in the following equation [9].

$$\frac{1}{C_{depl}^2} = \frac{2\left(V_{bi} - V_{appl} - \frac{2kT}{q}\right)}{q\varepsilon N} \qquad (1.27)$$

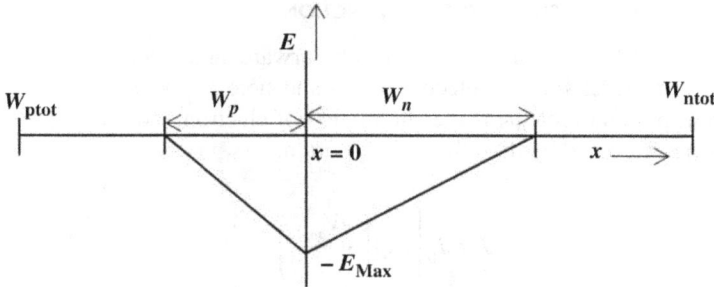

FIGURE 1.5 Electric field distribution in a semiconductor *p-n* homojunction.

Symbol N represents the concentration of impurity present in the substrate formed by the *p-n* junction. Equation (1.27) shows that the intercept of the ordinate in the $\dfrac{1}{C^2_{depl}}$ vs. V_{appl} plot gives the built-in potential at the junction. Existence of depleting capacitance can be sensed at the junction, which is un-biased or reverse biased. However, under forward biasing a different capacitance originated by minority carrier rearrangement across the junction dominates the former one. This capacitance is termed as diffusion capacitance as it originates by diffusion or injection of minority carriers (i.e., electron in *n*-type and hole in *p*-type semiconductor) across the junction. Small AC signal has a great effect on diffusion capacitance. At low frequency of associated AC signal the capacitance is proportional to the product of diffusion length and minority carrier concentration. So, the greater the injection of the minority carrier, the larger the diffusion capacitance.

(e) **Debye length (L_{Debye}):** Doping ofsemiconductor has a strange effect on the potential inside the depletion region and *C–V* characteristics. This effect works only if the distance of observation is less than a specific value. This characteristic limiting value of length that describes the change of potential and capacitance in the depletion region with respect to the doping profile is known as Debye length. The length can be calculated by using the following relation.

$$L_{Debye} = \sqrt{\frac{\varepsilon kT}{q^2 N}} \tag{1.28}$$

The concept reveals the fact that within Debye length the potential gradient is zero irrespective of doping profile. Thus, it can be concluded that the width of depletion region W must be more than L_{Debye} so that the Poisson equation is valid. Fortunately, the width of the depletion region at thermal equilibrium is eight and ten times that of L_{Debye} for silicon and GaAs, respectively [9]. Equation (1.28) signifies that the increase of N decreases L_{Debye}. Temperature has a direct effect on this length as well.

1.2.7 CURRENT DENSITY IN THE P-N JUNCTION

Under electrical biasing of V_{appl} volt (both forward and reverse) or illumination, common Fermi level splits to electron (E_{Fn}) and hole (E_{Fp})quasi-Fermi levels. The total current per unit area (J) is the summation of electron and hole current density. The J can be expressed by the following equation.

$$J = J_0 \left[exp \left(\frac{qV_{appl}}{kT} \right) - 1 \right]$$ (1.29)

The detailed derivation is discussed in section 3.1. In the equation, the relationship, $qV_{appl} = E_{Fn} - E_{Fp}$ holds to be true within the depletion region and the saturation current density is expressed as below.

$$J_0 = qn_i^2 \left[\frac{D_{ele}}{L_N N_{ACCP}} + \frac{D_{hol}}{L_P N_{DON}} \right]$$ (1.30)

where L_P and L_N are the diffusion length of the hole and electron inside the material. The first term of equation (1.30) is the contribution of diffusion of electrons inside the p-side of the junction. On the other hand, the second term is due to diffusion of holes in the n-side of the junction. The J increases with V_{appl} linearly after exceeding $3kT/q$. Conversely, under reverse bias at the initial biasing the current density increases slightly and then gets saturated at a level of $-J_0$ at large reverse biasing.

1.2.8 P-N HETEROJUNCTION

The p-side and n-side of this kind of junction are composed of different materials with suitable work functions. The principle of drift-diffusion for electrons and holes in such a junction is like that of the homojunction. However, there is a CB and VB discontinuity in the heterojunction due to the difference in electron affinity of the two materials involved. The Fermi energy levels of the two materials are aligned and a space charge region is formed around the junction. Accordingly, band banding arises in the region of depleted charge. However, part of the semiconductor far away from the junction acts as a normal piece semiconductor. The energy band alignment in the depletion region of a heterojunction is categorized into three types as depicted in Figure 1.6 (a)–(c).

(a) Type-I heterojunction: This type of heterojunction is formed between the two materials if the band gap of one material is less than the other. A necessary criterion of this type is that E_c and E_v of the lesser band-gap material completely shrinks inside the band gap of the larger band-gap material. Transportation in such a type of heterojunction demands energy equivalent to the corresponding band offset, i.e., CB offset (ΔE_{CON}) and VB offset (ΔE_{VAL}) for electron and hole inside the lower band-gap material, respectively. A typical example of this kind is a heterojunction between AlGaAs and GaAs.

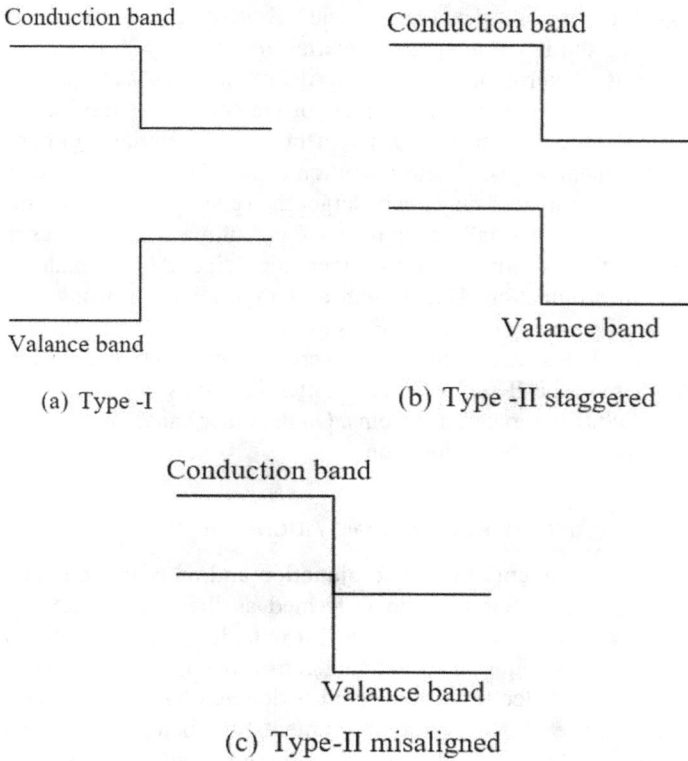

FIGURE 1.6 Different types of heterojunction configurations: (a) Type-I; (b) type-II staggered and (c) type-II misaligned.

(b) Type-II heterojunction: If the E_c and E_v of a material (material I say) is at higher energy position than that of different material (material II) after the Fermi level alignment, a type-II heterojunction is formed. The ΔE_{CON} occurred at the heterojunction may or may not be larger than the energy band gap of material I. If the offset ΔE_{CON} is smaller than the band gap of material I, a staggered type-II heterojunction is formed. On the other side, larger ΔE_{CON} creates misaligned type-II heterojunction. Under such a configuration electrons in material II need to gain energy equal to ΔE_{CON} to move to material I. In contrast, holes in material II lose ΔE_{VAL} energy to overcome the junction and reach material I. Staggered type-II semiconductor heterojunctions are observed in ZnO/Si and AlGaSb/InAs material systems. On the other hand, GaSb/InAs exhibits a type-II misaligned heterojunction.

(c) Type-I heterojunction: This type of heterojunction is formed between the two materials if the band gap of one material is less than the other. A necessary criterion of this type is that E_c and E_v of lesser band-gap material completely shrink inside the band gap of the larger band-gap material. Transportation in such a type of heterojunction demands energy equivalent to corresponding band offset, i.e., CB offset (ΔE_{CON}) and VB offset (ΔE_{VAL}) for the electron and hole inside the lower band-gap material, respectively. A typical example of this kind is a heterojunction between AlGaAs and GaAs.

(d) Type-II heterojunction: If the E_c and E_v of a material (material I say) is at higher energy position than that of different material (material II) after the Fermi level alignment, a type-II heterojunction is formed. The ΔE_{CON} occurred at the heterojunction may or may not be larger than energy band gap of material I. If the offset ΔE_{CON} is smaller than the band gap of material I, a staggered type-II heterojunction is formed. On the other side, larger ΔE_{CON} creates misaligned type-II heterojunction. Under such a configuration electrons in material II need to gain energy equal to ΔE_{CON} to move to material I. In contrast, holes in material II lose ΔE_{VAL} energy to overcome the junction and reach material I. Staggered type-II semiconductor heterojunctions are observed in ZnO/Si and AlGaSb/InAs material systems. On the other hand, GaSb/InAs exhibits a type-II misaligned heterojunction.

1.2.9 ELECTRON AFFINITY RULE OF BAND ALIGNMENT

The energy band alignment at the heterojunction and on both sides of it requires a reference line. This reference line is termed as the vacuum level of energy. The level is the energy of a free electron just outside the semiconductor surface. The amount of energy required to lift an electron from the minima of CB to the vacuum level is called electron affinity and is denoted by χ. The electron affinity rule defines the nature of energy band alignments in a heterojunction considering electron affinities of the materials involved. According to the rule, the difference in electron affinities, i.e., $|\chi_I - \chi_{II}|$ equalsis equal to the conduction band offset ΔE_{CON} at the interface. Assuming that the sum of CB and VB offsets is equal to the difference in energy band gap of the two materials forming the heterojunction, the valence band offset is given by, $\Delta E_{VAL} = \Delta E_g - \Delta E_{CON}$ where ΔE_g is the difference in energy gap of the two materials. However, this rule is based on mechanisms that occur in the bulk of the semiconductor irrespective of any exact reason in the heterointerface. Thus, the all-time validity of the rule is not guaranteed, therefore experimental ways to determine the band offset are adopted in practice. According to the sign convention positive $E_{cn} - E_{cp}$ is interpreted as positive ΔE_c while positive $E_{vp} - E_{vn}$ means ΔE_v is positive and vice versa. In the expressions, suffix n and p stand for n-type and p-type semiconductors, respectively, and where the quantity is evaluated.

1.3 JUNCTION CHARACTERIZATION PARAMETERS OF THE HETEROJUNCTION

Poisson's equation is the basis on which device characterization and the following parameters are derived. These parameters define the nature of the device and are used to calculate the performance of parameters. Here only the expression for parameters is presented. The derivation of the expressions is not presented in this book.

(a) **Built-in potential:** The built-in potential across the depletion region is given by,

$$qV_{bi} = E_{gp} - \Delta E_C + kT \ ln \left[\frac{N_{ACCP} N_{DON}}{N_{cn} N_{vp}} \right]$$ (1.31)

N_{cn} and N_{vp} r, respectively, are the effective conduction band and valence band DOS where suffix represents the type of semiconductor it belongs to.

(b) **Depletion with:** Width towards n-side of the junction:

$$W_n = \sqrt{\frac{2\varepsilon_n \varepsilon_p N_{DON} V_{bi}}{q\left(\varepsilon_p N_{ACCP} + \varepsilon_n N_{DON}\right) N_{ACCP}}}$$ (1.32)

Width towards p-side of the junction:

$$W_p = \sqrt{\frac{2\varepsilon_n \varepsilon_p N_{ACCP} V_{bi}}{q\left(\varepsilon_p N_{ACCP} + \varepsilon_n N_{DON}\right) N_{DON}}}$$ (1.33)

Net width of deletion region is the summation of the two partial widths that occurs on both sides of the junction as expressed below.

$$W = W_n + W_p = \sqrt{\frac{2\varepsilon_n \varepsilon_p \left(N_{DON} + N_{ACCP}\right)^2 V_{bi}}{q\left(\varepsilon_p N_{ACCP} + \varepsilon_n N_{DON}\right)\left(N_{DON} N_{ACCP}\right)}}$$ (1.34)

where, ε_n and ε_p are the refractive index of n-type and p-type materials used.

(c) **Depletion capacitance:** The junction capacitance under thermal equilibrium of the heterojunction (C_{HET}) can be indicated as follows [4]:

$$\frac{1}{C_{HET}^2} = \frac{2\left(\varepsilon_p N_{ACCP} + \varepsilon_n N_{DON}\right) V_{bi}}{q\varepsilon_n \varepsilon_p N_{DON} N_{ACCP}}$$ (1.35)

(d) **Current density:** Carrier transport through depletion region in the heterojunction cannot be structured simply by the drift-diffusion model as the band spikes originating at the interface due to CB and VB discontinuities causes an additional barrier. However, by assuming smooth transition of ΔE_c and ΔE_v as in the case of graded junction, a simple drift-diffusion model can be implemented for the heterojunction as well. The net current passing through the structure is dependent on material and applied bias as showcased in equation 1.36. In fact, the saturation current density in this case is different from that of the homojunction as given below.

$$J_0 = \left[\frac{qD_{elep} n_{ip}^2}{L_{Np} N_{ACCP}} + \frac{qD_{holn} n_{in}^2}{L_{Pn} N_{DON}} \right] \tag{1.36}$$

In this equation the terms n_{ip} and n_{in} represent intrinsic carrier concentration in the material on the p-side and n-side, respectively, of the heterojunction. A similar analogy stands for diffusion constant and diffusion length in the equation. So far, we have discussed the electrical parameters and underlying working principles of homo and hetero p-n junction semiconductor materials. However, in practice, not all such calculations and expression are exact until specified assumptions are made. The analysis and characterization of current transport in p-n junction using the drift-diffusion model considers the following assumptions:

- Electrons inside the depletion region are assumed to be at equilibrium with the electrons in n-side of the junction. The Fermi level is the same inside and outside of the depletion region in the$n$-side of the junction.
- Holes inside the depletion region are assumed to be at equilibrium with the holes in p-side of the junction. The Fermi level is the same inside and outside of the depletion region in the p-side of the junction.
- As the Fermi level and band edge difference is increasing inside the depletion region, majority carrier concentration decreases on both sides of the junction inside the depletion region.
- Depletion approximation assumes that majority carrier concentration inside the depletion region is zero on both sides of the junction.

1.4 CARRIER DYNAMICS INSIDE THE SEMICONDUCTOR

The electrons resting inside the CB are bounded by an energy and any mechanism triggering loss of the energy lead those electrons to combine with holes resting in the VB. In some cases, the energy loss due to recombination is converted to a bunch of photons known as light and this type of recombination is called radiative. In others the loss of energy is spent on fueling other electrons in the CB and the recombination is called Auger. The later type of recombination is also known as non-radiative recombination. However, the phenomenon where CB electrons jump to the VB and recombine holes is not alone. In most of the materials one more commonly occurring recombination phenomenon exists and is known as Shockley-Read-Hall (SRH). This is characterized by recombination of free electrons and holes at some impurity centers located in the forbidden gap of the material. The Auger and SRH recombination are non-radiative in nature. In all those mechanisms the rate at which the charge carriers recombine is dependent on its concentration. The rate of recombination of radiative, Auger and SRH are discussed below. Recombination of charge carriers in a semiconductor takes place in several ways as discussed in the following section. All the recombination rates are expressed under the assumption of low-level injection, i.e., excess carriers ($\Delta n \approx \Delta p$) are smaller than majority carrier concentration. Mathematically, the status of carrier concentrations assumed is $\Delta n \ll n$ or $\Delta p \ll p$.

1.4.1 RECOMBINATION OF CHARGE CARRIERS

(a) **Band-to-band radiative:** In this process an electron in the CB jumps back to VB and recombines with a hole present there. Direct band-gap semiconductors, especially III-V compounds, have higher probability of undergoing the radiative process. The electron jumped from the CB loses its energy in the form of a photon, thus the name radiative recombination is entitled to the process. The rate of recombination (R_{rad}) is dependent on concentrations of both types of carriers.

$$R_{rad} = C_{rad}np \tag{1.37}$$

The symbol C_{rad} is the radiative recombination coefficient, which is much larger ($\approx 10^{-10}$ cm^{3}s^{-1}) for direct band-gap than for indirect band-gap semiconductors ($\approx 10^{-15}$ cm^{3}s^{-1}). The time interval for which an electron remains free before recombining is known as its lifetime. In this context electron lifetime in p-type semiconductor ($\tau_{lif\_e\_bb}$) and hole lifetime in n-type semiconductor ($\tau_{lif\_h\_bb}$) are written as below.

$$\tau_{lif\_e\_bb} = \frac{1}{C_{rad}N_{ACCP}} \tag{1.38}$$

$$\text{and } \tau_{lif\_h\_bb} = \frac{1}{C_{rad}N_{DON}} \tag{1.39}$$

(b) **Shockley-Read-Hall (SRH) recombination:** Carrier recombination, primarily in indirect band-gap materials like germanium, silicon etc. takes place at bulk trap states located inside the bandgap. The recombination rate in the process (R_{srh}) depends on some major parameters like electron and hole capture cross sections denoted by σ_n and σ_p, respectively, trap density in the bulk (N_{trap}) and the product of carrier concentrations np.

$$R_{srh} = \frac{\sigma_n \sigma_p N_{trap} v_{th} \left(pn - n_i^2 \right)}{\sigma_n \left(n + n_i \right) + \sigma_p \left(p + n_i \right)} \tag{1.40}$$

Thermal velocity is denoted by v_{th} in the equation. It is to be noted that only those trap states at the vicinity of mid-gap are effective provider of recombination centers. The rate equation presented above is for one trap level only. Corresponding electron lifetime in p-type semiconductor ($\tau_{lif\_e\_srh}$) and hole lifetime in n-type semiconductor ($\tau_{lif\_h\_srh}$), respectively, are as follows:

$$\tau_{lif\_e\_srh} = \frac{1}{\sigma_n v_{th} N_{trap}} \tag{1.41}$$

$$\text{and } \tau_{lif\_h\_srh} = \frac{1}{\sigma_p v_{th} N_{trap}} \tag{1.42}$$

(c) Auger recombination: An electronon recombination with a hole gives rise to energy but not in the form of a photon. Instead, the released energy is obtained by a second electron, which collides with crystal and releases the energy to another electron. The additional energy is given up through sequential collision via several electrons. The rate of recombination (R_{Aug}) in this type of process is proportional to carrier concentration.

$$R_{Aug\_e} = C_{Aug\_e} n^2 p; \text{ for electron.} \tag{1.43}$$

$$R_{Aug\_h} = C_{Aug\_h} p^2 n; \text{ for hole.} \tag{1.44}$$

The electron lifetime ($\tau_{lif\_e\_Aug}$) and hole lifetime ($\tau_{lif\_h\_Aug}$) can be written as below.

$$\tau_{lif\_e\_Aug} = \frac{1}{C_{Aug\_e} N_{DON}^2} \tag{1.45}$$

$$\text{and } \tau_{lif\_h\_Aug} = \frac{1}{C_{Aug\_h} N_{ACCP}^2} \tag{1.46}$$

All three types of recombination mechanisms are depicted in Figure 1.7.

(b) Shockley-Read-Hall (SRH) recombination: Carrier recombination, primarily in indirect band-gap materials like germanium, silicon etc. takes place at bulk trap states located inside the bandgap. The recombination rate in the process (R_{srh}) depends on some major parameters like electron and hole capture cross sections denoted by σ_n and σ_p, respectively, trap density in the bulk (N_{trap}) and the product of carrier concentrations np.

(c) Auger recombination: An electronon recombination with a hole gives rise to energy but not in the form of a photon. Instead, the released energy is obtained by a second electron, which collides with crystal and releases the energy to another electron. The additional energy is given up through sequential collision via several electrons. The rate of recombination (R_{Aug}) in this type of process is proportional to carrier concentration.

1.4.2 SEMICONDUCTOR UNDER EXCITATION

In the previous section, the rate of electron and hole recombination is discussed and indicates reduction of n and p with respect to time. However, in practice, the

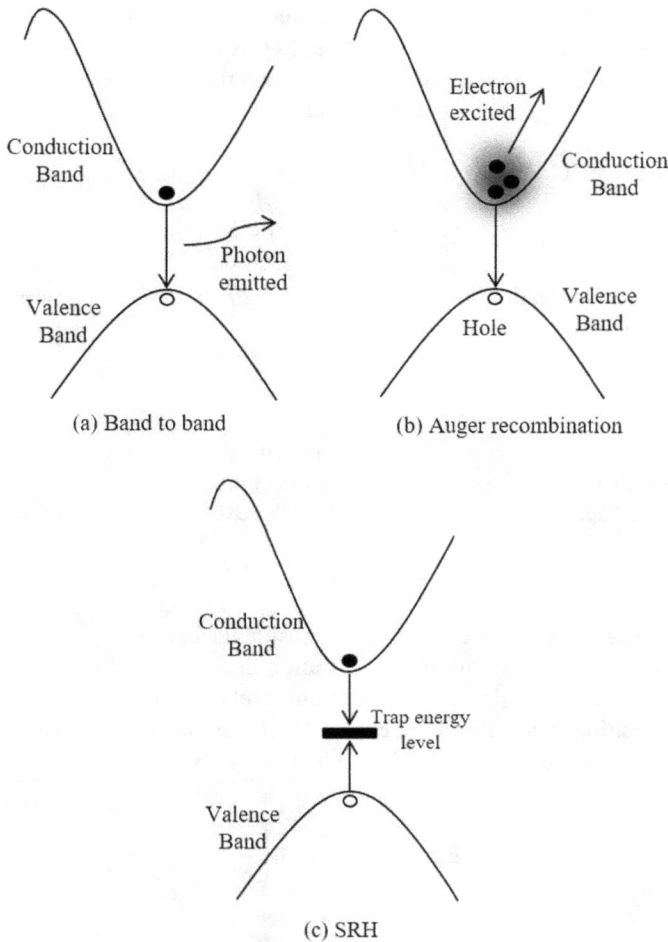

FIGURE 1.7 Carrier recombination mechanisms in semiconductor: (a) Band to band;
(b) Auger and (c) Shockley-Read-Hall recombination.

concentration of electrons and holes remain constant irrespective of time if no excitation is applied. Under equilibrium condition, the electron and hole concentration are
governed by the law of mass action as given below.

$$pn = n_i^2 \tag{1.47}$$

This indicates the loss of carrier concentrations due to recombination is compensated
by another process known as generation. Under equilibrium condition, the rate of
carrier generation (G) is equal to the net recombination rate (R) i.e., $G = R$. If any
excitation such as applied voltage or exposer to light are imposed, semiconductors
undergo a change in minority carrier concentration due to the generation of carriers.
Generation of carriers does not affect noticeably the majority carrier concentration in

semiconductors. However, minority carriers due to their small concentration in the material, senses the effect of such a change. Let us assume the change in minority electron and hole concentrations are Δn and Δp at a particular instant of time and the radiative recombination rate equation becomes [1]

$$R_{BE} = C_{rad}(p + \Delta p(t))\,(n + \Delta n(t)) \tag{1.48}$$

$$= R_{rad} + C_{rad}\,(p\,\Delta n(t) + n\,\Delta p(t)) + C_{rad}\Delta p(t)\,\Delta n(t) \tag{1.49}$$

Under low level injection $\Delta n(t) = \Delta p(t)$ and $C_{rad}\Delta p(t)\,\Delta n(t)$ is negligibly small. So, the equation becomes

$$R_{BE} = R_{rad} + C_{rad}\,(p + n)\,\Delta n(t) \tag{1.50}$$

Excitation leads generation of charge carriers inside the crystal as well. The rate of generation of the charge carriers depends on the type of excitation and ambient conditions. In general, the rate of generation under excitation can be written as below.

$$G_E = G + G_e \tag{1.51}$$

where G_e is the generation of excess carriers inside the crystal.

The mass action law of carrier concentration does not hold true in such a scenario. Both electrons and holes possess different levels of Fermi energy in the material under this condition. The splitting of energy levels produces two quasi-Fermi energy levels for electrons and holes as expressed by the following equations.

$$E_{fn} = E_{Fin} + kT\,ln\left(\frac{n}{n_i}\right) \tag{1.52}$$

$$E_{fp} = E_{Fin} - kT\,ln\left(\frac{p}{n_i}\right) \tag{1.53}$$

The carrier concentrations in the material are related by the expression depicted below.

$$\frac{pn}{n_i^2} = exp\left[\frac{E_{fn} - E_{fp}}{kT}\right] \tag{1.54}$$

Under light illumination on the semiconductor, pairs of electrons and holes are generated by a rate G_e thereby disturbing the system from thermal equilibrium ($pn \neq n_i^2$). To restore its equilibrium, electrons and holes get recombined by a rate R in several ways. The continuity equation interprets total carrier concentration change over the region of interest in terms of the difference in generation and recombination rate and drift-diffusion current through the region as given by equations in one dimension below.

$$\frac{\partial n}{\partial t} = G - R + \frac{1}{q}\frac{\partial J_{ele}}{\partial x} \tag{1.55}$$

$$\frac{\partial p}{\partial t} = G - R - \frac{1}{q}\frac{\partial J_{hol}}{\partial x} \tag{1.56}$$

Equations (1.55) and (1.56) are valid for electron and hole transportation, respectively.

1.4.3 GENERATION RATE

Generation of charge carriers in a piece of semiconductor while exposed to light is dependent on position and wavelength of the photon. Generation rate denoted by G of electrons and holes is proportional to the intensity gradient of light in the semiconductor material. An empirical expression for G in a semiconductor exposed by a light intensity is given below.

The position-dependent light intensity inside the semiconductor material is denoted by $P(x)$ and is related to externally exposed (normal incidence) intensity P_0 as given below.

$$P(x) = P_0 (1 - R) exp(-\alpha x) \tag{1.57}$$

Absorption coefficient (α) and reflectance (R) at the front surface are material-dependent parameters.

1.5 LIGHT EMISSION FROM LED

The working mechanism of LEDs can be explained with the fundamental physics of semiconductors as discussed in the previous sections. The radiative recombination can be obtained from different structures, which are discussed in this section. Based on the structure the LEDs are classified as a homojunction and a double heterojunction or a quantum well (QW).

1.5.1 HOMOJUNCTION LED

The foremost LED is nothing but a *p-n* junction made on direct band-gap semiconductor materials such as GaAs, InAs, InP etc. As discussed in section 1.2.5, a *p-n* homojunction forms a depletion region at the vicinity of the junction where no free charge carriers exist. Moreover, a built-in potential blocks the diffusion of electrons from *n*-side to *p*-side. Similarly, the same potential hinders hole diffusion from *p*-side to *n*-side. The current through the junction (J) is the summation of electron (J_{ele}) and hole (J_{hol}) current components if no recombination current is assumed to be exists inside the depletion region. These currents are constituted by drift and diffusion elements as given below.

$$J_{ele} = -qn\mu_n E + qD_n \frac{dn}{dx} \tag{1.58}$$

$$J_{hol} = -qp\mu_p E + qD_p \frac{dp}{dx} \tag{1.59}$$

where, q, E, D and μ are charge, electric field across the depletion region, diffusion constant and mobility of the charge carrier, respectively. The suffixes n and p represent the parameter is for the electron and hole, respectively. The first and second term in equations (1.58) and (1.59) represent drift and diffusion currents, respectively. It is noteworthy that the current due to recombination inside the depletion region is assumed to be null in the derivation, which is not negligible in a practical scenario.

Under forward biasing the built-in potential is reduced and the diffusion of charge carriers are enabled. Consequently, a small number of holes can move from p-side to n-side inside the VB as depicted in Figure 1.8. Similarly, a small number of electrons move from n-side to p-side inside the CB. As mentioned earlier the total current across the depletion region is summation of drift and diffusion components but later is not affected under moderate biasing (V). The reason being sufficient electric field inside the thin depletion region for velocity saturation of the charge carriers. The

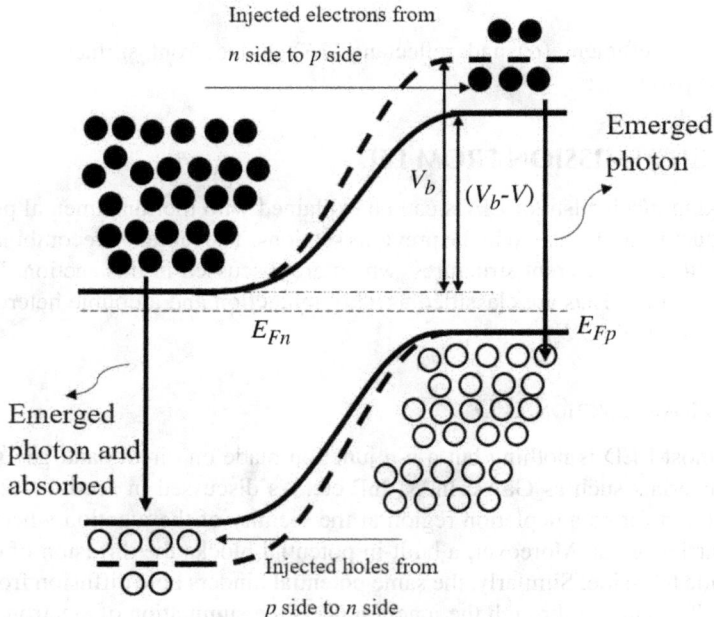

FIGURE 1.8 Energy band diagram of a p-n junction under forward biasing.

drift current inside the depletion region is negligible under forward biasing and is not considered for calculation of J in this chapter. The other component is diffusion current at the vicinity of depletion region, which greatly affects the value of J.

To understand this, let us assume that uniform concentration of the electron away from and at the edge of the depletion region in p-side are n_p and n_{wp}, respectively. Similarly, p_N and p_{WN} are hole concentration away and at the edge of the depletion region in n-side, respectively. The following relations hold true [9].

$$n_{WP} = n_p exp(qV/kT) \tag{1.60}$$

$$p_{WN} = p_N exp(qV/kT) \tag{1.61}$$

The excess charge carrier concentration is the difference between concentration under equilibrium and after excitation. The excess electron concentration at the edge of the depletion region inside p-side is

$$\Delta_{nP} = \left(exp\left(\frac{qV}{2kT}\right) - 1 \right) nP \tag{1.62}$$

The excess hole concentration at the edge of the depletion region inside n-side is

$$\Delta pN = \left(exp\left(\frac{qV}{2kT}\right) - 1 \right) pN \tag{1.63}$$

The electron concentration profile inside the p-region is given by the following equation.

$$n_p(x) = \Delta n_p \, exp([x+w_p]/L_N); \; x < \text{depletion edge in } p\text{-side } (-w_p) \tag{1.64}$$

Similarly, the hole concentration profile inside the n-region is given by

$$p_N(x) = \Delta p_N \, exp(-[x-w_N]/L_p); \; x > \text{depletion edge in } n\text{-side } (w_N) \tag{1.65}$$

From equations (1.64) and (1.65) the diffusion current can be derived using the following relationship.

$$\text{For electron: } J_{diff\_n} qD_n \frac{dn_p(x)}{dx} \bigg|_{x=-w_P} \tag{1.66}$$

$$\text{For hole: } J_{diff\_p} = qD_p \frac{dp_N(x)}{dx} \bigg|_{x=w_N} \tag{1.67}$$

The diffusion current is evaluated at the edges of the depletion region. The charge carrier concentration profile and corresponding diffusion current distribution is depicted in Figure 1.9(a). Equations (1.64) and (1.65) vividly indicate decreased nature of the injected minority carrier concentration far away from the depletion edge. This is due

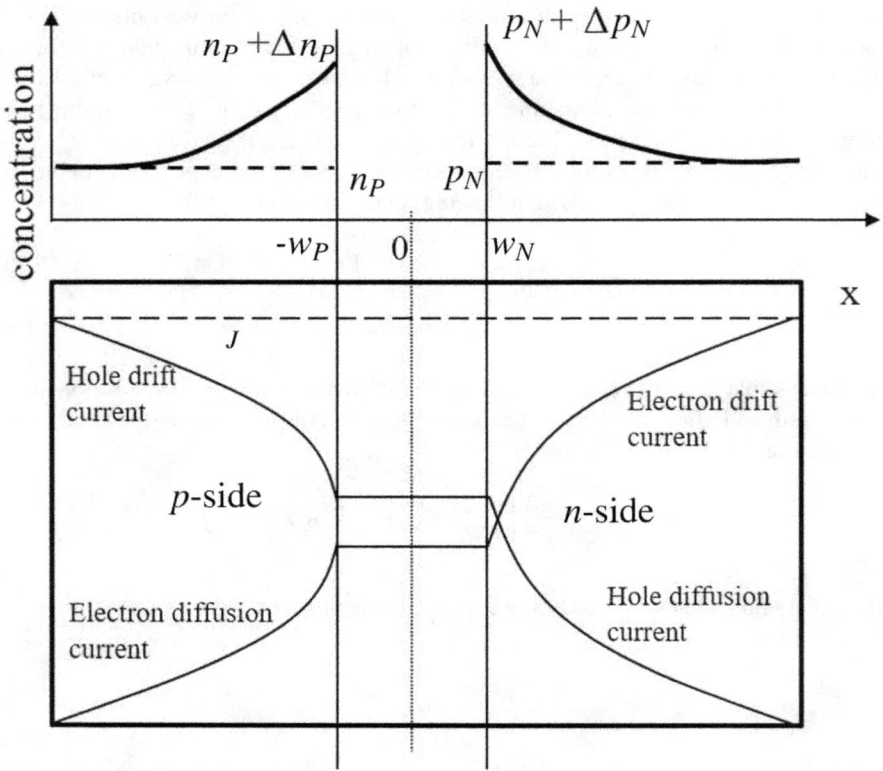

FIGURE 1.9 Elements of current density across the *p-n* junction.

to the recombination of minority with majority charge carriers inside the bulk region. The number of recombined majority charge carriers are compensated from the bulk itself, which in turn provides a drift current as depicted in Figure 1.9(a). The total current J is the summation of diffusion (given by equations (1.66) and (1.67)) and drift current inside the bulk region of *n*- and *p*-sides. The final expression for J is shown in equation (1.29).

If we assume trap-mediated recombination current inside the depletion region, an additional current, as given below should also be considered during the calculation of total current J.

$$J_{depl} = \frac{qn_iW}{2\tau} exp\left(\frac{qV}{2kT} - 1\right) \tag{1.68}$$

The recombination of charge carriers inside the *n*- and *p*-side can enable emission of light out from the bifacial of the junction. However, in practice either side of the junction is made thinner so that the light can emit from that side without being absorbed.

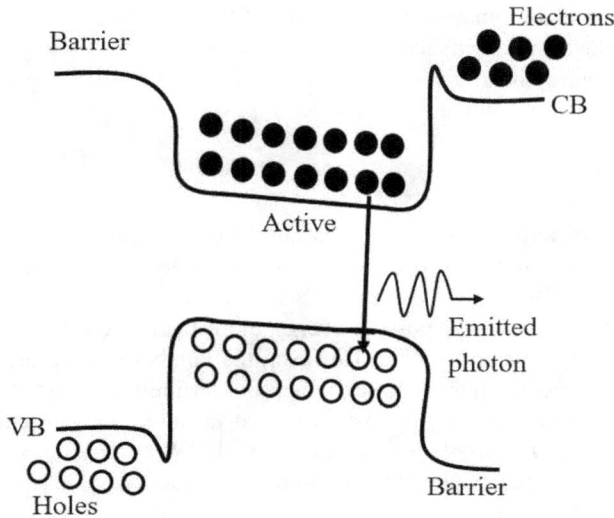

FIGURE 1.10 Schematic of energy band diagram of QW made from double heterojunction.

Shortcomings of homojunction LED:

(i)   Photons emitted inside the emitter layer are reabsorbed. As a result, emission volume of the photon must be close to the outer surface of the emitter layer.

(ii)   The surface of a material bears enormous defect encouraging non-radiative recombination.

(iii)   The minority carriers injected into the emitter layer has long diffusion length. Consequently, the emission volume of the photon is enlarged.

1.5.2 QUANTUM WELL LED

To overcome the shortcomings of conventional LED, QW is introduced. The QW is a well of energy formed using double heterojunction between two different materials as depicted in Figure 1.10.

The concept of QW LED portrays confinement of electrons and holes inside a potential well and allowing it to recombine radiatively. The QW consists of an active region made of low direct band-gap material such as GaAs. The active region is sandwiched by two barrier layers made of high band-gap materials such as Al_xGa_{1-x} As. The n- and p-type of barrier layers contribute electron and hole to the potential well, respectively. These charge carriers are confined and can easily recombine radiatively. The advantages of QW LED are mentioned below.

(a)   The volume of emission can be easily controlled by managing the thickness of the potential well.

(b) Radiative recombination at the surface of the material does not play any role on the device performance.

(c) Larger energy band gap of barrier layers allows the photon generated due to recombination inside the active layer to pass through.

Shortcomings of QW LED:

(a) **Carrier leakage:** As mentioned earlier the charge carriers are confined in the active region by barrier energy (ΔE) on both sides of the QW. The electrons' distribution inside the active region are governed by Fermi–Dirac distribution. Some of the electrons possess energy more than ΔE, which are capable of leaking out the active region. However, the number of such electrons is low. Radiative recombination of such electrons inside the barrier layer provides less emission efficiency compared to that inside the active region. The carrier leakage is a problem in a fundamental QW LED. The number of such electrons can be calculated by the following equation.

$$n_{ae} = \int_{E_{CB}}^{\infty} D_{OS} f_E(E) dE \tag{1.69}$$

where, D_{OS} and $f_E(E)$ represent the density of energy states and Fermi–Dirac function of the active layer, respectively. The corresponding current due to electron leakage (J_L) can be approximated by the following equation.

$$J_{depl} = J_L = -qD_n \frac{dn_{ae}(x)}{dx} \tag{1.70}$$

where, $n_{ae}(x)$ is the profile of electron inside the p-type barrier layer and $x = 0$ corresponds to the junction between active and p-type barrier layer. The negative sign in the above equation represents the loss mechanism.

(b) **Carrier overflow:** In the QW LED structure the charge carriers are injected from the barrier layers to the active layer by means of external applied voltage. The higher the applied current corresponding to that external biasing, the more is the injection quantity. This indicates a linear increase of emitted light intensity with respect to applied current. However, the increase in the charge carrier concentration boosts up the Fermi energy level inside the active region. Consequently, some of the carriers overflow and move to the barrier layer even if high external biasing is applied. Under that condition, the intensity of emitted light gets saturated. The current corresponding carrier overflow is directly proportional to the width of the active region. In addition, it depends on the effective density of state and ΔE of the material system.

1.5.3   EFFECT OF THICKNESS AND DOPING IN QW LED

The thickness and doping of the active layer play an important role in designing a QW-based LED. In most of the LEDs a moderate thickness of the active layer is desirable.

A thick active layer would become the same as homojunction LED and the advantage of QW is missed. On the other hand, a thin active region encourages charge carrier overflow and degrades the quantum efficiency of the device. The active region is generally undoped but moderate doping (less than that of barrier) of the order of 10^{16} cm^{-3} is used in some of the LEDs [2]. The preferred type of doping active region is acceptor atoms. As electrons have more mobility than that of holes donor doping is seldom used.

1.5.4 PERFORMANCE PARAMETERS OF LED

The major operation of an LED is to convert injected electrons to photons. In ideal LEDs, various loss mechanisms are not considered but in a real scenario these cannot be neglected. The performance parameters of LEDs provide a measure of the degree of performance under different non-ideal conditions. These are presented in Table 1.1.

1.6 OPTICAL PHYSICS OF LED

The physics behind LEDs is governed by optical and electrical mechanisms. In the previous sections, electrical background is discussed and the optical part is described in this section.

1.6.1 PHOTON SPECTRUM OF IDEAL LED

A LED emits non-coherent light, which consists of a narrow band of frequency. The maximum energy of the emitted light from the LED is nearly equal to the energy band gap of the material. For the ideal case, this is $0.5kT + E_g$. The momentum of electrons and holes should be the same for a radiative recombination to occur. Assuming electron distribution inside the allowed energy bands of the material is governed by Boltzmann's equation, the spectrum of light emitted from the LED is depicted in Figure 1.11. The intensity of light is proportional to the density of state of the material and the charge carrier distribution. The full width at half maximum (FWHM) of the emitted spectrum of a L ED is approximately 1.8 kT.

TABLE 1.1
Different Performance Measuring Parameters of LED

Parameter	Formula
Extraction efficiency	$\eta_{ex} = \dfrac{photon\,emission\,rate\,to\,free\,space}{photon\,emission\,rate\,from\,active\,region}$
EQE	$\eta_{eqe} = \dfrac{photon\,emission\,rate\,to\,free\,space}{electron\,injection\,rate\,to\,active\,region}$
IQE	$\eta_{iqe} = \dfrac{photon\,emission\,rate\,from\,active\,region}{electron\,injection\,rate\,to\,active\,region}$
Power efficiency	$\eta_{po} = \dfrac{output\,optical\,power}{input\,electrical\,power}$

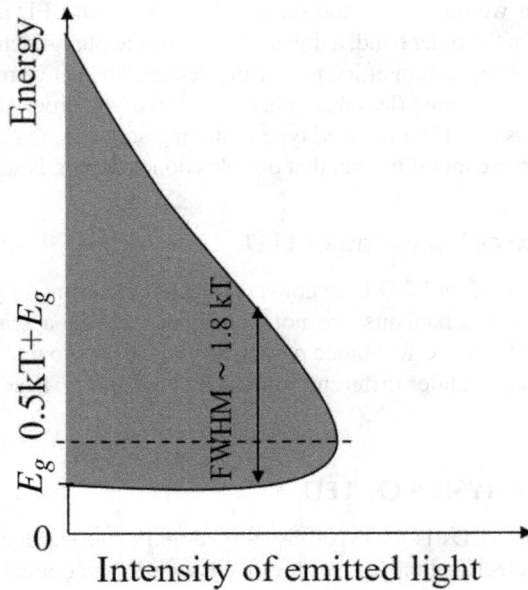

FIGURE 1.11   Ideal spectrum of light emitted from LED.

1.6.2 EMISSION CONE AND RADIATION OF LED

LED emits light from the region where charge carrier recombination takes place. Such regions are undoubtedly near to the surface of the LED. The surface of the device makes a junction with air and the higher refractive index of the semiconducting material (n_{sem}) than that of the air (n_{ar}), total internal reflection takes place. Consequently, a fraction of the total photon emitted inside the semiconductor comes out of the device. This fraction is described by *emission cone*. The critical angle at the semiconductor–air interface decides the *emission cone* of the LED. The fraction of optical power that is emitted out of the device with respect to total power generated is approximated as given below.

$$\frac{P_{out}}{P_{tot}} \approx 0.25 \left(\frac{n_{ar}}{n_{sem}} \right)^2 \tag{1.71}$$

In a typical LED let us assume that light is radiated from source at an angle of α and refracted at an angle of β as depicted in Figure 1.12. The surface area of the emission cone is a part of spherical geometry with radius r. The intensity of light escaped to the air is given by the following relationship.

$$I_{ar} = \frac{P_{tot}}{4\pi r^2} \left(\frac{n_{ar}}{n_{sem}} \right)^2 \cos(\beta) \tag{1.72}$$

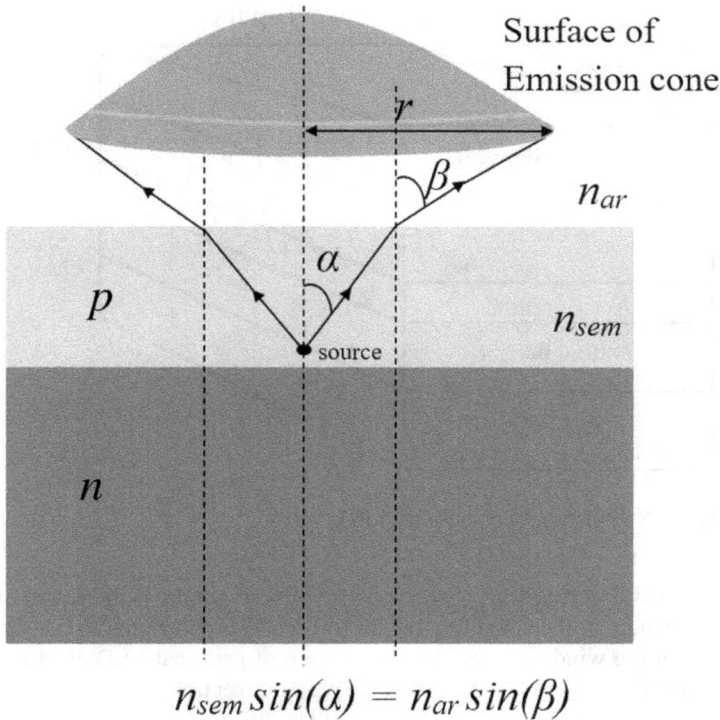

$$n_{sem}\, sin(\alpha) = n_{ar}\, sin(\beta)$$

FIGURE 1.12   Pathway of emitted light from LED.

The distribution of the intensity of light is a function of refracted angle and is popularly known as the Lambertian pattern of radiation. This pattern is different for different types of surfaces of LED. Hemispherical and parabolic LED surfaces provide an isotropic and highly directed pattern of radiations.

1.6.3   SOME PRACTICAL LED STRUCTURES

The practical LED structures are designed to improve IQE and extraction efficiency. Firstly, we discuss the design to improve the former quantity. The most practical LED is made from multiple QW structure in which a thin active layer is sandwiched between barrier layers. The metal contacts are deposited on the barrier layers. However, the injected current through the metal contact at the top is confined within a small area of the barrier layer. Consequently, the carrier injected area inside the active region is small and selected area-based emission is enabled. To overcome such a problem a thin layer of a material is deposited on the top barrier layer and is known as current spreading or window layer. The overall structure of the LED is fabricated on a suitable substrate as depicted in Figure 1.13. The solid and dashed arrowhead lines indicate injected current with and without a current spreading layer in the LED structure, respectively. The shape of metal contact and thickness of such a layer affects

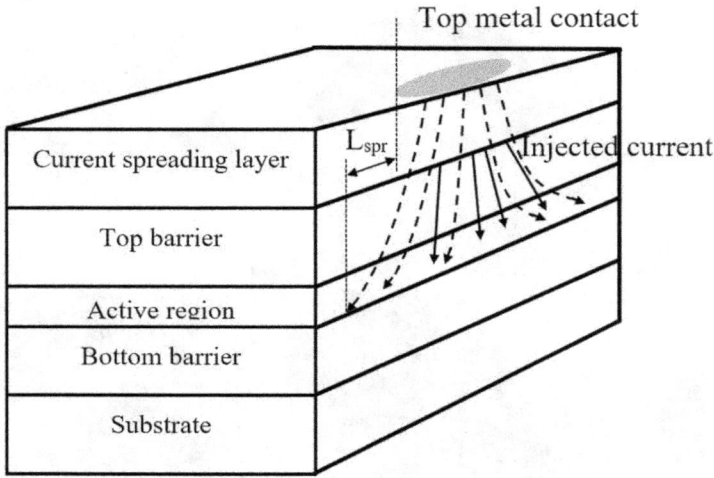

FIGURE 1.13   Schematic of a typical QW LED.

the current spreading length (L_{sprd}) enormously. For example, if the shape of the top contact is rectangular, the spreading length is directly proportional to the square root of thickness of the window layer. Another design of practical LED aims to improve the extraction efficiency. As discussed in the previous section, the light emission cone of LED limits the extraction efficiency. To widen the surface area of light emission, different geometries of LED structure are designed. Example cone shaped, spherical, truncated inverted pyramid, and pedestal-shaped LEDs [4].

The UV emitted LED is quite a demanding device and development of such technology is a trend. In the existing literature, it is found that the wavelength range of emission of AlGaN LED covers up the entire UV wavelength range. This is possible by controlling the alloying of gallium nitride (GaN) and aluminiumnitride (AlN) in AlGaN [10]. However, several challenges need to be rectified in this technology such as high defect densities (threading dislocation densities and point defects), lack of effective thermal management, challenging strategies for efficient carrier injection and light extraction. Over the past few years, several research groups across the globe have put their valuable effort to solve these challenges and the corresponding strategies and techniques have been reported. Lee et al. have demonstrated high-performance AlGaN-technology-based deep UV LEDs by applying periodic air-voids-incorporated nanoscale patterns using nanosphere lithography and epitaxial layer overgrowth on sapphire substrate [11]. Jain et al. have demonstrated migration-enhanced lateral epitaxial overgrowth methodology for the growth of thick films of AlN and AlGaN on trenched AlGaN/sapphire substrate with low-defects, which in turn causes significant improvement in device lifetime [12]. Miyake et al. has demonstrated that the film uniformity and crystallinity of the AlN buffer layer can be substantially improved by annealing in the atmosphere of carbon-saturated N_2-CO mixture at relatively high temperature [13]. Hirayama et al. have used ammonia (NH_3) pulse-flow multilayer (ML) growth technique to obtain the AlN layer with low threading dislocation density; which in turn offered high-performance AlGaN- and InAlGaN-based deep ultraviolet LEDs [14]. Murotani et al. have demonstrated the IQE of Si-doped AlGaN QW can be significantly enhanced by using silicon-doping [15]. Takano et al. have comprehensively discussed the influence of several features such as a transparent AlGaN:Mg contact layer, a Rh mirror electrode, an AlN template on a patterned sapphire substrate, and encapsulation resin on the light extraction efficiency and hence to obtain a high-performance AlGaN-based UV LEDs [16].

Apart from group-III nitride technology, ZnO has also been explored significantly in the existing literature for the advancement of LED application. Various key advantages of ZnO such as cost-effectiveness, non-toxicity, excellent chemical stability, and ease of synthesis and preparation; makes it a preferable and potential candidate over GaN technology for LED applications. In the existing reports, ZnO has been significantly explored for homojunction- and heterojunction-based LEDs for UV and visible region emission spectra. A schematic diagram of a blue emission LED based on the work of Jeong et al. is presented in Figure 1.14 (a) [17]. The device structure consists of an n-type Al-doped ZnO, nanowire shaped ZnO and p-type Mg-doped GaN deposited on a sapphire substrate using MOCVD and RF-magnetron sputtering. A InGaN multiple QW LED emitting blue light was proposed in the work of Akita et al. [18]. The device consists of p-type GaN contact, p-type AlGaN electron blocking layer, InGaN multiple QW, n-type GaN and sapphire substrate. A device schematic based on their work is shown in Figure 1.14 (b).

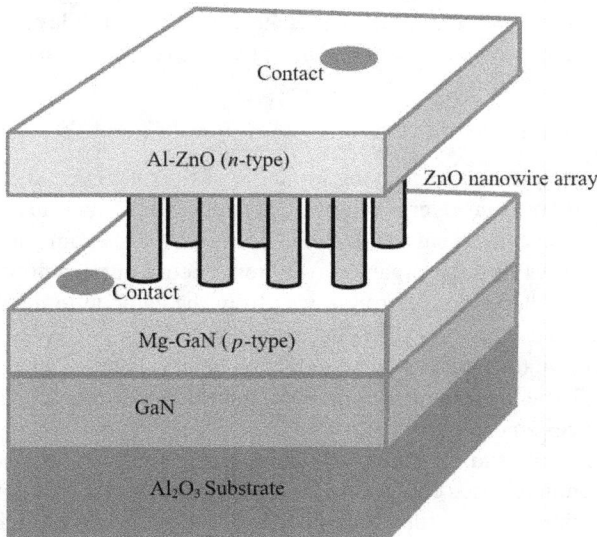

FIGURE 1.14 Schematic diagram of (a) Al-ZnO/ZnO nanowire/Mg-GaN LED and (b) InGaN multiple QW LED.

Source: [17, 18].

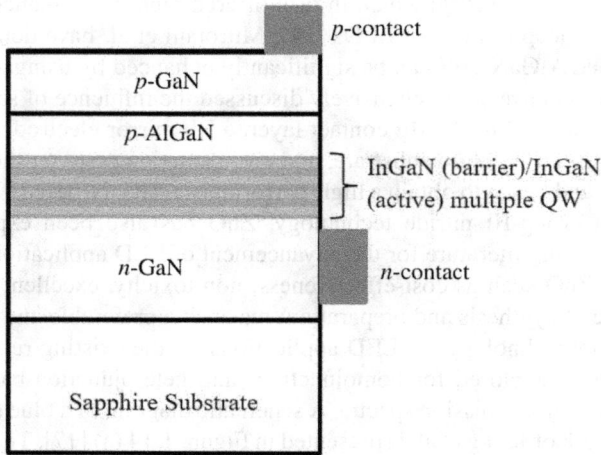

FIGURE 1.14   (Continued)

1.7   INTRODUCTION TO ORGANIC LED (OLED)

It needs to be mentioned that the inorganic LED has some structural disadvantages. The thin organic layers help in more significant light emission. Also, OLED does not require glass support unlike the conventional inorganic technology-based LCD and LED. Moreover, plastic material can be used in the OLED device for fine display applications. Moreover, the organic layer has a distinct advantage over the inorganic layer because it does not require lattice matching. Organic multilayers, such as electron (ETL) and hole transport layer (HTL), and an active layer, make up OLED.

1.7.1   OLED DEVICE STRUCTURE

Electro luminescence is the underlying principle of the OLED device structure. The electroluminescent organic layer is sandwiched between ETL and HTL, emitting light in response to electrical currents as depicted in Figure 1.15. From cathode to anode, electric current flows through organic layers (an electric current flow of electrons). Under a DC bias voltage, an electron travels from the cathode to the emissive layer (EML) via ETL, and a hole travels from the anode to EML via HTL. The electrons discover the holes in the emitting and conduction layers. There is an accumulation of electrons and holes at the transport layer/EML interface. The interface under external biasing experiences an energy barrier.

The generated hole and electron recombined in the emissive layer and formed an exciton. The exciton releases energy as a light photon and emits light [19]. The chemical structure of the light-emitting material affects the colour of the light. Because of this, material design can influence it in any way it wishes. Intensity or luminosity is directly proportional to the amount of electricity flowing through it. This exciton is called a coulombically bound electron-hole pair in EML. These electrons and holes recombine in the active layer and form an exciton.

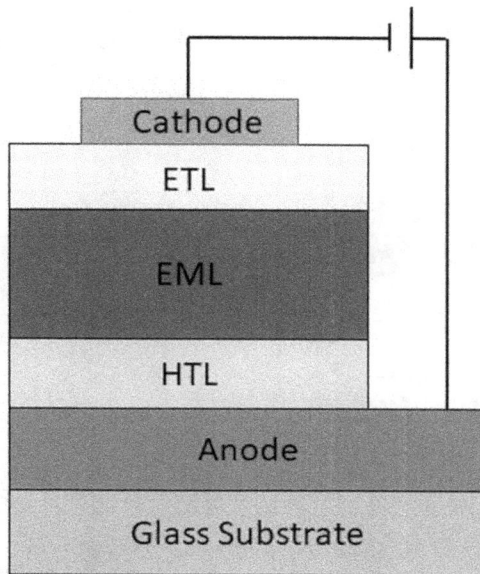

FIGURE 1.15 Schematic illustrations of the OLED device.

1.7.2 EXCITON DECAY MECHANISM

The exciton formed in the active layer contains two unpaired electrons in different orbitals, each with a different spin (triplet and singlet excitons). It is possible to divide OLEDs into fluorescent and phosphorescent OLEDs based on the emitter material, as both electrons and holes are fermions with a half-integral spin. There are singlet excitons (S=0) and triplet excitons (S=1) depending on the spin combination, which determines the wavefunction configuration. The singlet excitons (S=0) result from spin combinations opposite the spin combination (S=1). According to Figure 1.16, singlet states have a 25% chance of occurring, while triplet states have a 75% chance of happening. Fluorescent emission refers to the heat lost from the remaining 75% of these excitons that could not generate light in the past. On the other hand, a significant advance by researchers from Princeton University and Southern California expressed that 100% of excitons could be converted into light by electro phosphorescence mechanism, now also known as phosphorescence. Therefore, achieving up to four times the efficiency of a conventional fluorescent OLED with a phosphorescent one is possible. Light is produced due to the decay of excitons, which results in spontaneous emission. Light emission from a triplet exciton is uncommon. Triplet exciton has a very low probability of occurring because of the forbidden transition, which is not the case for most materials. So, the triplet exciton has lower energy than the singlet exciton because of the repulsive nature of the spin-spin interaction between electrons with the same spin [20,21]. Electroluminescence can be generated in an organic layer by decaying radiatively (calculating the total number of electron-generated photons).

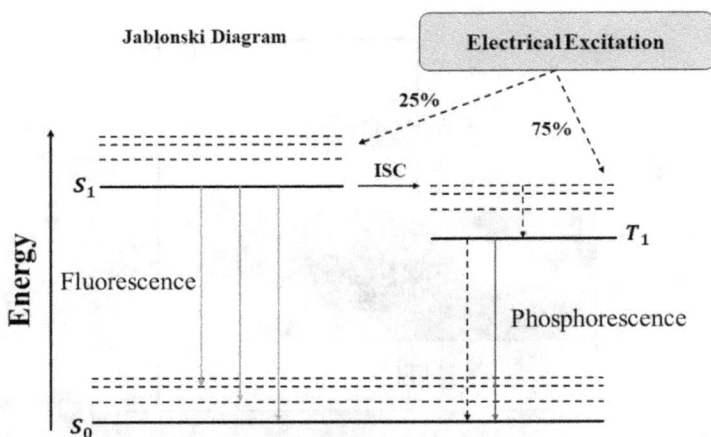

FIGURE 1.16 The schematic Jablonski diagram displays the decaying state formed by fluorescence and phosphorescence.

The OLED device's external quantum efficiency [22, 23]. Exciton decay occurs quickly and easily in the singlet state, while in the triplet state, it is more complex [24]. Transitions that are not radiative, on the other hand, can be caused by a wide range of factors such as internal conversion, vibrational relaxation and intersystem crossing are all examples of non-radiative transitions that do not use radiation to make the transitions. In other words, by combining the S1 and T1 states, decay in triplet excitons can be permitted, and excited electrons can transition to the lower triplet excited state. After a singlet (S1-fluorescent) or a triplet (T1-phosphorescent) transition, depending on the emitter material, these decay and emit light, which is associated with the energy difference between HOMO and LUMO.

1.7.3 Elements of the OLED Structure

As shown in previous section, OLED consists of HTL, ETL, EML, cathode and anode. While its energy band alignments depend on material electron affinity, band gap and ionization potential. The low electron affinities of the organic layer serve as hole-transporting materials and the high electron affinities and ionization potential serve as electron-transporting materials. Therefore, the detailed description of different layers is defined below [24–28],

HTL: In OLEDs, electron-donated materials are used as hole-transporting materials. Allows for the injection of holes from the anode, accepts holes, and transfers the injected holes to the emissive layer. Additionally, the HTL acts as an electron blocking layer, preventing the release of the electron to the EML from anode side. The typical hole transport layers are NPB (*N*, *N*'-bis(naphthalen-1-yl)-*N*, *N*'-bis(phenyl)-benzidine), TAPC (4,4'-cyclohexylidenebis [N, N-bis(4-methyl phenyl) benzenamine]) etc. Its chemical structures are shown in Figure 1.17.

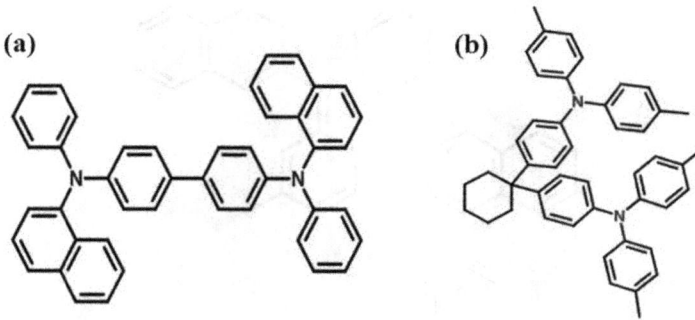

FIGURE 1.17   Representative chemical structure of (a) HTL-NPB and (b) HTL-TAPC.

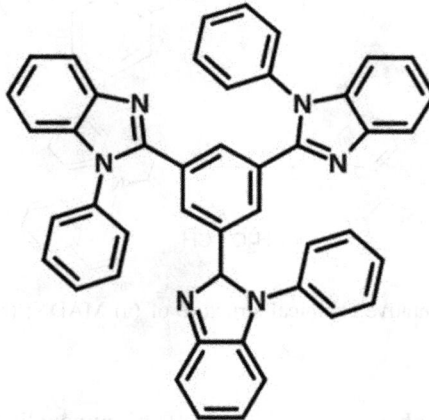

FIGURE 1.18   The chemical structure of ETL-TPBi.

ETL: In OLEDs, the electron transporting material is an electron-accepting material. The electron transport layer (ETL) accepts an electron from the cathode. It transfers the injected electron to the emissive layer, facilitating electron injection. Notably, the electron-transport layer also serves as a hole blocking layer. The most common example of ETL is TPBi (2,2′,2″-(1,3,5-Benzinetriyl)-tris(1-phenyl-1-H-benzimidazole)). The chemical structure of TPBi is shown in Figure 1.18.

EML: The EML or the active layer under the response of electric current, emits light. The typical fluorescent emission layers are a well known green emitter, Tris(8-hydroxyquinoline)aluminum (Alq3), MADN (2-methyl-9,10-bis(naphthalen-2-yl) anthracene), BEPP (bis(2-(2-hydroxyphenyl)-pyridine) beryllium), and phosphorescent layer namely α-NPD: Ir (MDQ)$_2$ as EML1 [N, N′ Di(1-naphthyl)- N, N′-diphenyl-(1,1′-biphenyl)-4,4′-diamine doped with

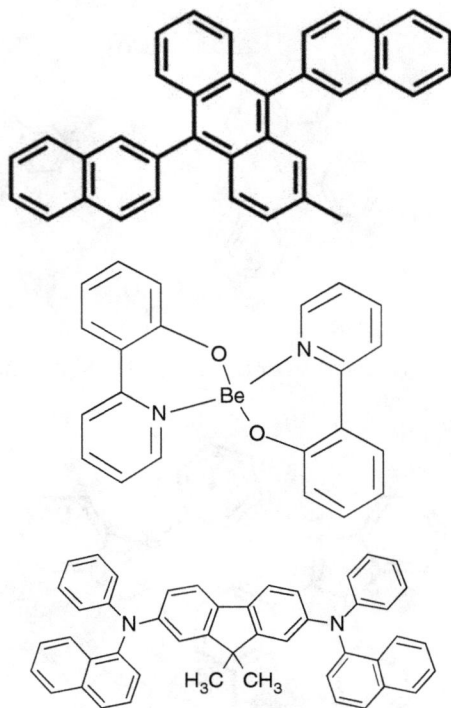

FIGURE 1.19 Representative chemical structure of (a) MADN; (b) BEPP; (c) α-NPD: Ir (MDQ)$_2$.

the red-emitting dye bis (2-methyldibenzo[f, h] quinoxaline)(acetylacetonate) iridium(III)], α -NPD, as a host material and Ir(MDQ)$_2$ as a guest material. TCTA: Ir(ppy)$_3$ as EML2 [tris(4-carbazoyl-9-ylphenyl) amine] doped with the green-emitting dye tris [2-phenylpyridinato-C2, N] iridium (III)]. In this TCTA, as a host material and Ir(ppy)$_3$ as a guest material. The chemical structure is shown in Figure 1.19.

1.7.3.1 Electrodes

- **Anode:** ITO acts as an anode (the anode removes electrons (adds the electrons to the "holes") when a current passes through the device) due to high reflectivity, good electrical conductivity and high work function, and easy patterning ability.
- **Cathode:** The cathode (depending on the OLED type, it may or may not be transparent) injects electrons when a current passes through the device. The most commonly used cathodes are aluminum (Al) and silver (Ag) because they are more durable and low-work function metals.

1.7.4 PURPOSE OF HIL AND EIL

These layers are selected based on their orbital levels, namely HOMO for HIL and LUMO for EIL. Additionally, the work function of the respective electrodes is matched to the orbital levels of the individual layers to select the optimal layers. The charge injection should be improved to bring these two levels closer together. As the energy barrier between the transport layers and the electrode work function is lowered, a more significant number of charge carriers can access the transport layers that correspond to them through these layers [29, 30].

1.8 OLED DEGRADATION MECHANISMS

Reliability is a key parameter to commercialization of organic devices. Extensive research has focused on small-molecule OLED degradation. Instability in operation and storage reduces device efficiency. Long-term intrinsic decay causes operational instability in OLED devices that results in reduced device efficiency. It causes two things: (a) luminance decay is initially rapid, then slower; (b) operating voltage increases gradually at a constant current [31–36].

OLED degeneration has several causes.

(a) **Anode contact:** Considerably high energy barrier at the anode contact produces large joule heat, causing local molecule aggregation. Plasma treatments improve ITO anode contact. Enhanced hole-injection improves OLEDs' voltage, efficiency and reliability. Hole-injecting layers can be used to modify ITO anodes.

(b) **Crystallization:** One of the most important mechanisms for organic thin-film degradation. This is because organic thin films made by vapour evaporation are amorphous and glassy. Because the glass transition temperatures of most HTLs are relatively low (Tg).

(c) **Non-emissive sites:** Since moisture and other environmental factors affect the OLED device performance, encapsulation of these devices is very important. Non-emissive spots grow over time. Before device construction, ITO/glass may have a particle or asperity, or by depositing an organic layer using vacuum deposition. Both particle types may be larger than the organic layer, causing shadowing during deposition. Water and oxygen can enter gaps in coverage. Since dark spots form when metal oxide and hydroxide grow at cathode/organic interfaces, their growth rates depend on the metals' chemical stability. Thicker metal reduces dark spot density.

1.9 OLED DEPOSITION TECHNIQUES

OLEDs are deposited using a solution process or dry deposition method. Small molecular organic material deposition uses dry techniques like vacuum thermal evaporation and organic vapour phase deposition. Dry techniques make it easy to deposit large areas of uniform and homogenous film, and small aromatic molecules are too soluble for solution processing of thick films. Spin coating, ink jet printing and contact stamping are commonly used solution processing methods for polymer organic materials. Due to their large molecular weight vacuum thermal evaporation is not feasible. Their evaporation temperature is higher than their decomposition temperature. Spin coating and vacuum thermal evaporation are described below [37, 38].

- **Spin coating:** Spin coating dissolves a solid (usually a polymer) in an organic solvent. This solution is placed on a substrate, allowed to wet the coating area, and then spun at 1000–10000 rpm. The spinning centrifugal force expels most of the solution, but some remains. Liquid film thickness is controlled by adhesion forces at the substrate/liquid interface, solution viscosity, and air/liquid friction. Low-pressure solvents can form an indefinite thin liquid film on spinning substrates. Typical organic solvents used in the spin coating have a high vapour pressure and evaporate quickly in unsaturated environments. This thin liquid film eventually thins (1–60s), leaving behind a thinner (1–1000nm), flat solid film of the initial solvated liquid. The patterned deposition is not possible with spin coating.
- **Vacuum thermal evaporation (VTE):** The simplest deposition method is VTE. Electric current heats a resistive metal boat (usually tungsten, molybdenum or tantalum). The boat's organic material evaporates when heated. The evaporation occurs in a low-pressure (10^{-6} torr) vacuum chamber to keep the evaporated material as pure as possible. Molecules are released from the source with constant speed and direction that changes only if encounters a cold substrate or chamber surface. Due to the even distribution of initial molecular trajectories, the deposited film has uniform thickness irrespective of any incident angle. 70–99% of the boat's material is deposited on the vacuum chamber walls depending on throw distance. Shadow masking is used to pattern VTE materials. Shadow masking patterns material by preventing deposition in unwanted areas. Shadow masks are made from thin metal foil with perforations.

1.10 OLED DEVICE IMPLICATIONS

A high-efficiency OLED device with better external quantum efficiency is needed in the present scenario. According to the literature, charge imbalance, outcoupling efficiency, and longer lifetime affect device performance. Different ways have been found to solve the efficiency problem. First, energy consumption should be reduced by improving energy efficiency. Secondly, charge imbalance and enhanced hole injection rate should be reduced using a hole modulating layer between HTL/EML or EML/ETL. Thirdly, using a microlens array (MLAs) on the planar OLED's glass substrate improves the efficiency and spectral shift of the device. However, the combination of fluorescence and phosphorescent emission layer intensifies the exciton operation. Many obstacles must be overcome to realize this technology's potential fully [39–42]. This group includes the following:

- **Light extraction:** With the current planar OLED device structures, most light is trapped inside the substrate due to total internal refraction (TIR) at the glass/air interface. Since MLAs do not alter the device architecture or functionality, they are being suggested to counter light extraction efficiency by using an array of MLAs onto a glass substrate.
- **Charge imbalance:** It is the excess number of non-recombined charges or charge imbalance that does not recombine all-electron and holes to form exciton in the active layer degrade device efficiency due to the charge imbalance. Based on spin configuration and energy level of the organic material, these excitons produce photons in either a singlet or a triplet transition. As a result, the difference in energy levels between HOMO and LUMO is linked to the colour of light that can be achieved by adjusting organic materials. The balanced injection of carriers into the EML can be performed by inserting a hole-modulating layer (HML) between the EML/ETL or the HTL/EML. This change will increase hole injection and reduce charge imbalance in the recombination zone.
- **Fine patterns with intense colours:** It are much more difficult for blue and red devices to achieve the same efficiency level due to the human perception of light intensity peaking in green. Despite significant advancements in active organic products, better blue, green and red emitters are still required by competitor technologies.
- **Device stability:** OLEDs have a short lifespan. These devices may be particularly vulnerable to the effects of humidity and heat. Small weight, slim profile and flexibility are still challenging to maintain despite encapsulation's ability to lessen the impact of hostile environments. The device's performance cannot degrade significantly over time, even if stored or operated for an extended period.

1.11 TYPES OF OLED DEVICE

Based on the architecture of the device different types of OLED are observed. These are discussed below.

- **Bottom or top emission:** It refers to the direction in which emitted light exits the device, not to the direction of the OLED display. If the light is emitted from the substrate on which the OLED is fabricated, it is called bottom emission device. The light emitted by an OLED device is graded based on whether or not it exits through a lid added after the device was made. Using top-emitting OLEDs in active-matrix applications is more convenient because they are easier to integrate with a non-transparent transistor backplane. Bottom emitting AMOLEDs are typically made with a non-transparent TFT array attached to the bottom substrate, resulting in significant transmit light blockage.
- **Transparent OLEDs:** Displays that emit light from the top and bottom can be made with transparent organic light-emitting diodes (OLEDs), which have transparent or semi-transparent contacts on both sides of the device (transparent).

TOLEDs have the potential to significantly improve contrast, making it much easier to read displays even when exposed to direct sunlight. This technology could be used in head-up displays, smart windows and augmented reality applications.

- **Graded heterojunction:** Gradually reducing the electron-hole to electron transporting chemical ratio is the goal of graded heterojunction OLEDs. This results in a quantum efficiency nearly double that of current OLEDs.
- **Stacked OLEDs:** There are many advantages to using an OLED panel that stacks the sub-pixels on top of each other rather than one another, such as an increase in gamut and colour depth and a reduction in pixel gaps. RGB (and RGBW) pixels are mapped next to each other in most other display technologies, which reduces resolution.
- **Inverted OLED:** Inverted OLEDs use a bottom cathode connected to the drain end of an n-channel TFT, particularly for the low-cost amorphous silicon TFT backplane used in manufacturing AMOLED displays.

1.12 MODELLING APPROACH OF AN OLED DEVICE

The fundamental structure of an OLED device consists of an active layer material sandwiched between ETL and HTL. It emits light in response to electric current. 1D drift-diffusion, charge carrier transportation equations, first-generation model (characterized by sharp energy states and use the classical Einstein relation to link diffusion constant and charge mobility) and second-generation models (ECDM/ EGDM foresee disordered, spread energy levels and need to account for a generalized Einstein relation) used to study the physics of OLED devices. This can also be implemented using software such as SETFOS to study OLED electrical and optical properties. The following equations are described below [43–46]:

$$J_e(x) = e\mu_e(x,E)n(x)E(x) + D(\mu)\frac{dn(x)}{dx} \tag{1.73}$$

$$\frac{dE(x)}{dx} = \frac{e}{\epsilon\epsilon_0}\big[p(x) - n(x)\big] \tag{1.74}$$

$$\frac{\partial n(x)}{\partial t} = \frac{1}{e}\frac{\partial J_e(x)}{\partial x} - r(x)p(x)n(x) \tag{1.75}$$

With electronic charge e, free space permittivity ϵ_0, medium permittivity ϵ, J_e indicates net electron current, p holes density, n electrons density, recombination rate coefficient $r(x)$ and fickian diffusion constant $D(\mu)$. Nevertheless, local charge mobilities affect the rate of recombination in organic semiconductors. The Langevin recombination efficiency parameter is described below.

$$R = \eta np\big(\mu_n + \mu_p\big)\frac{q}{\epsilon} \tag{1.76}$$

Using Maxwell Boltzmann statistics, the electron and hole concentrations are expressed by

$$n = N_{LUMO} exp\left(\frac{F_n - E_{LUMO}}{kT}\right) \quad (1.77)$$

$$p = N_{HUMO} exp\left(\frac{E_{HUMO} - F_p}{kT}\right) \quad (1.78)$$

E_{LUMO} and E_{HUMO} are the energy level of an organic layer related to LUMO and HUMO. While the quasi-Fermi levels are F_n and F_p.

1.12.1 SMALL SIGNAL ANALYSIS (IMPEDANCE SPECTROSCOPY)

Impedance spectroscopy, a sophisticated experimental technique for characterizing organic devices, uses a sinusoidal voltage signal to measure the phase-shifted periodic current response. The current implementation in SETFOS has some restrictions on the device complexity that can be accommodated. Still, it allows bipolar transport and multiple charge mobility models (constant, Poole-Frenkel, EGDM and ECDM). But it only covers single-layer devices (no heterojunction) in EGDM or ECDM mobility models. A sinusoidal voltage bias is first applied to the device to briefly summarize this approach. Then the out-of-phase sinusoidal current is measured:

$$V = V_0 + V_{ac}e^{j\omega t} \quad (1.79)$$

$$J = J_0 + J_{ac}e^{j\omega t} \quad (1.80)$$

Typical outputs of these simulations are the admittance (Y), the impedance (Z), the capacitance (C) and the conductance (G), defined as follows:

$$Y = \frac{J_{ac}}{V_{ac}} = G + j\omega C \quad (1.81)$$

$$Z = \frac{V_{ac}}{J_{ac}} \quad (1.82)$$

Suppose the applied sinusoidal voltage is small enough. The device response is linear, and the output current will also be a sinusoidal function. This approximation can then be used to efficiently solve the coupled system of equations describing the device operation made of the continuity and Poisson equations.

1.12.2 OPTICAL ANTENNA MODEL FOR OLED: PHOTON EXTRACTION

Photons are produced by OLEDs due to the radiative decay of excitons. However, because of the cavity effect, extracting photons from the devices is more difficult because most OLED devices are constructed using multiple layers. Concurrently, organic layers and electrodes have a shorter thickness than the wavelength of visible light, ranging from 10 to 200 nanometers (380–780 nm). Meanwhile, one of the electrodes is frequently made of reflective material to enhance the light collection on the other side. Therefore, the light emission of OLED devices will be modulated due to multiple reflections and transmissions within the device.

Meanwhile, the cavity structure affects the rate at which the excitons decay radiatively. The angular dependence of the emission characteristics in OLED is caused by the cavity effect. Especially when applying OLED technology to display panels, the luminance drops and colour shifts at high viewing angles are significant. People can improve the colour quality and increase the outcoupling efficiency of the device by tuning the structure. They also quantitatively calculated the dipole emission in the cavity structure. First, a quick introduction to the model will be given. The light mode decomposition in OLEDs, the comparison between weak and strong cavity OLEDs, and the layer thickness dependence in OLEDs are tested here. Among the other cavity effects examined is the light mode decomposition in OLEDs.

1.12.3 Mode Analysis of Multilayer Device Structures

Chance et al. included electromagnetic radiation at the vicinity of conductive surface to model fluorescence emission in optical cavities[44]. In this model, molecules are thought of as driven harmonic dipole oscillators that have had their oscillations dampened by a feedback mechanism. The following equation explains the oscillating dipole moment $\vec{p}$ dynamics:

$$\frac{d^2}{dt^2}\vec{p} + b_0\frac{d\vec{p}}{dt} + \omega^2\vec{p} = \frac{e^2}{m}\vec{E}_R(\omega) \tag{1.83}$$

with $\vec{E}_R(\omega)$ the interface-reflected field at dipole position, m the effective mass of the dipole, e the elementary charge, and ω as the oscillator frequency in the undamped case. The radiative power of a perpendicular dipole in an infinite birefringent (IB) medium having an ordinary refractive index n_{or} and extraordinary refractive index n_{ex} is

$$q_0 b_o(perpendicular) = \frac{|p|^2 \omega K_0^3}{12\pi\varepsilon_o}n_{or} \tag{1.84}$$

where q_0 be the IQE of the dipole and $K_0 = \frac{2\pi}{\lambda}$. The equation for calculating the radiative power of a horizontal dipole in an IB medium with an ordinary refractive index n_{or} and an extraordinary refractive index n_{ex} is as follows:

$$q_0 b_o(parallel) = \frac{|p|^2 \omega K_0^3}{12\pi\varepsilon_o}n_{or}\left(\frac{3n_{or}^2 + n_{ex}^2}{4n_{or}^2}\right) \tag{1.85}$$

So, the effect of the optical environment over the radiative part is summarized as below,

$$b = q_0 b_o F + (1 - q_0) b_o \qquad (1.86)$$

where F is the Purcell factor that indicates the normalized emitted power. Equation (1.87) below determines the absorbance of each layer

$$F = \int_0^\infty f(u) du \qquad (1.88)$$

where $f(u)$ indicates the optical feedback and radiation at any given position. In addition, the relationship between intrinsic lifetime τ_0 and a total lifetime of a dipole τ is defined as

$$\frac{\tau_0}{\tau} = \frac{b}{b_0} = 1 - q_0 + q_0 F \qquad (1.89)$$

1.12.4 Luminance, Radiance and Emission Spectrum

The integral within an infinitesimal solid angle around θ is shown below,

$$R(\theta) = \int E_\theta(\lambda) d\lambda \qquad (1.90)$$

With

$$E_\theta(\lambda) = \frac{1}{2\pi} \cdot EmissionIntensity. \, E_\gamma . \Lambda(\lambda) . \Pi(\theta) \int_x \left(N_\gamma(x) . \, g_x(\theta, \lambda) \right) dx \qquad (1.91)$$

$$N_\gamma(x) = k_r(0) . \, S(x) \qquad (1.92)$$

With $k_r(0)$ the radiative decay and $S(x)$ exciton density distribution. The function $g_x(\theta, \lambda)$ determines the spectrum of the emitted photons out of the device, E_γ the photon energy, $\Lambda(\lambda)$ intrinsic luminescence spectrum. The projection function $\Pi(\theta)$ shows the apparent emissive surface. $\Pi(\theta)$ is defined as,

$$\Pi(\theta) = \begin{cases} 1, & when \ E_\theta \ is \ spectral \ radiance \\ \cos\theta, & when \ E_\theta \ is \ spectral \ intensity \end{cases} \qquad (1.93)$$

Luminance is expressed in cd/m² that measures the brightness of source. The following relation is defined as

$$L(\theta) = 683 \frac{lm}{W} \int \bar{y}(\lambda) E_\theta(\lambda) d\lambda \qquad (1.94)$$

Equation 1.94 represents the luminance of the source, $E_\theta(\lambda)$ denotes spectral emission, the photopic luminosity function is $\bar{y}(\lambda)$. The equation for external quantum efficiency is as follows:

$$EQE = \eta_{cb}\,\eta_{st}\,\eta_{rad}\,\eta_{out} \tag{1.95}$$

With η_{rad} radiative efficiency, η_{cb} is the balance of charge that depends on the injection of charge carrier, η_{out} is an outcoupling efficiency, η_{st} defined the fraction of exciton 0.75 for a triplet and 0.25 for a singlet.

1.12.5 COMMISSION INTERNATIONALE DE L'ECLAIRAGE

When it comes to colour perception and weighting functions (x, y, z), the CIE 1931 is used with a corresponding colour matching function defined as

$$X = \int_0^\infty S(\lambda)x'(\lambda)d\lambda \tag{1.96}$$

$$Y = \int_0^\infty S(\lambda)y'(\lambda)d\lambda \tag{1.97}$$

$$Z = \int_0^\infty S(\lambda)z'(\lambda)d\lambda \tag{1.98}$$

the tristimulus values are described below:

$$x = \frac{X}{X+Y+Z} \tag{1.99}$$

$$y = \frac{Y}{X+Y+Z} \tag{1.100}$$

Moreover, $z = 1 - x - y$, third chromaticity.

1.12.6 SOME RESULT OF OLED MODELS

To see the effect of the microlens array on top of the OLED as an external scattering structure, here we simulate the variation of EQE and extraction efficiency. Figure 1.20(a) depicts the external quantum efficiency of pyramid packing arrays on top of the OLED as an external scattering structure as a function of the integrated luminance. The figure reveals that the pyramid array has the highest EQE, at 36%, that is better than the previously reported EQE value in the literature, i.e., 34% [41]. While the roll-off occurred with a minimum EQE of 30% and reached a brightness level of 10^5 lm/m², it is essential to note that the roll-off is not linear. Despite this,

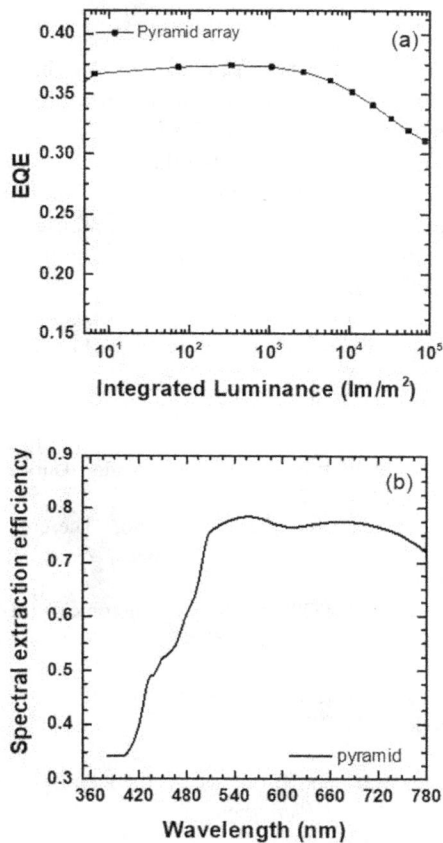

FIGURE 1.20 Comparison of EQE of pyramid microlens array onto a glass substrate of OLED device (a) EQE-integrated luminance characteristics; (b) spectral extraction efficiency-wavelength.

the MLAs 35% EQE kept at the power of luminance 10^3. Using a pyramid MLA on the emitting surface of the OLED substrate causes a significant increase in the spectral extraction efficiency. The variation of these efficiencies with wavelength is illustrated in Figure 1.20(b). The efficiency of spectral extraction for the pyramid packing type was measured at an average of 75%. The following considerations need to be taken into account. The micropattern was applied to the glass substrate in order to extract a larger number of photons that were primarily trapped in substate mode in conventional OLEDs as a result of the interaction between TIR and waveguiding at the boundaries of different refractive index materials. This was done because the micropattern does not cause any change in the device's architecture or functioning. Consequently, the OLED emits many rays in diverse orientation and have further polarization.

1.13 CONCLUSION

A thorough discussion on different types of LED is discussed in this chapter. The fundamental physics of the semiconductor such as energy band alignment of homojunction and heterojunction, drift diffusion of charge carriers across a *p-n* junction, light-semiconductor interaction has been discussed. Moreover, quantum well structure of LED and its different aspects are described briefly. Apart from LED, One of the most emerging technology is organic light emitting diode (OLED), so the detailed introduction of the OLED and its working principle is a part of the discussion in this chapter. The OLED is a highly attractive area of research at present for its low-cost fabrication and the impressive emission power. A generalized overview on working principle and status of OLED is also presented in the chapter.

REFERENCES

[1] E. F. Schubert, "Light-emitting diodes," Cambridge University Press, 2nd edition, ISBN-13 978-0-511-34476-3,2006.
[2] M. Kneissl, T.-Y. Seong, J. Han, and H. Amano, "The emergence and prospects of deep-ultraviolet light-emitting diode technologies," *Nature Photonics,* vol. 13, no. 4, pp. 233–244, 2019.
[3] D. Li, K. Jiang, X. Sun, and C. Guo, "AlGaN photonics: recent advances in materials and ultraviolet devices," *Advances in Optics and Photonics,* vol. 10, no. 1, pp. 43–110, 2018.
[4] Y.-S. Choi, J.-W. Kang, D.-K. Hwang, and S.-J. Park, "Recent advances in ZnO-based light-emitting diodes," *IEEE Transactions on Electron Devices,* vol. 57, no. 1, pp. 26–41, 2009.
[5] A. Khan, K. Balakrishnan, and T. Katona, "Ultraviolet light-emitting diodes based on group three nitrides," *Nature Photonics,* vol. 2, no. 2, pp. 77–84, 2008.
[6] J. Cho, J. H. Park, J. K. Kim, and E. F. Schubert, "White light-emitting diodes: history, progress, and future," *Laser & Photonics Reviews,* vol. 11, no. 2, p. 1600147, 2017.
[7] T. Kim et al., "Efficient and stable blue quantum dot light-emitting diode," *Nature,* vol. 586, no. 7829, pp. 385–389, 2020.
[8] Y. Kondo et al., "Narrowband deep-blue organic light-emitting diode featuring an organoboron-based emitter," *Nature Photonics,* vol. 13, no. 10, pp. 678–682, 2019.
[9] S. M. Sze and Kwok K. Ng, "Physics of semiconductor devices," John Wiley and sons, 3rd edition, p. 14, ISBN-I 3:978-0-471-14323-9, 2007.
[10] M. Kneissl, T-.Y. Seong, J. Han, and H. Amano. "The emergence and prospects of deep-ultraviolet light-emitting diode technologies," *Nature Photonics,* vol. 13, no. 4, pp. 233–244, 2019.
[11] D. Lee et al., "Improved performance of AlGaN-based deep ultraviolet light-emitting diodes with nano-patterned AlN/sapphire substrates," *Applied Physics Letters,* vol. 110, no. 19, p. 191103, 2017.
[12] R. Jain et al., "Migration enhanced lateral epitaxial overgrowth of AlN and AlGaN for high reliability deep ultraviolet light emitting diodes," *Applied Physics Letters,* vol. 93, no. 5, p. 051113, 2008.
[13] H. Miyake et al., "Annealing of an AlN buffer layer in N2–CO for growth of a high-quality AlN film on sap phire," *Applied Physics Express,* vol. 9, no. 2, p. 025501, 2016.

[14] H. Hirayama et al., "222–282 nm AlGaN and InAlGaN-based deep-UV LEDs fabricated on high-quality AlN on sapphire," *Physica Status Solidi (a)*, vol. 206, no. 6, pp. 1176–1182, 2009.

[15] H. Murotani, D. Akase, K. Anai, Y. Yamada, H. Miyake, and K. Hiramatsu, "Dependence of internal quantum efficiency on doping region and Si concentration in Al-rich AlGaN quantum wells," *Applied Physics Letters*, vol. 101, no. 4, p. 042110, 2012.

[16] T. Takano, T. Mino, J. Sakai, N. Noguchi, K. Tsubaki, and H. Hirayama, "Deep-ultraviolet light-emitting diodes with external quantum efficiency higher than 20% at 275 nm achieved by improving light-extraction efficiency," *Applied Physics Express*, vol. 10, no. 3, p. 031002, 2017.

[17] M.-.C. Jeong, B.-Y. Oh, M.-H. Ham, L. Sang-Won, and J-.M. Myoung. "ZnO-nanowire-inserted GaN/ZnO heterojunction light-emitting diodes," *Small*, vol. 3, no. 4, pp. 568–572, 2007.

[18] K. Akita, T. Kyono, Y. Yoshizumi, H. Kitabayashi, and K. Katayama. "Improvements of external quantum efficiency of InGaN-based blue light-emitting diodes at high current density using GaN substrates," *Journal of Applied Physics*, vol. 101, no. 3, p. 033104, 2007.

[19] N. Thejo Kalyani Hendrik Swart and S. J. Dhoble, Principles and Applications of organic light emitting diodes (OLEDs), Woodhead Publishing, 2017.

[20] M. A. Baldo, D. F. O'Brien, M. E. Thompson, and S. R. Forrest, "Excitonic singlet-triplet ratio in a semiconducting organic thin film," *Physical Review B*, vol. 60, no. 20, pp. 14422–14428, 11/15/ 1999.

[21] C. Rothe, S. King, and A. Monkman, "Direct measurement of the singlet generation yield in polymer light-emitting diodes," *Physical Review Letters*, vol. 97, no. 7, p. 076602, 2006.

[22] M. I. Mishchenko, L. D. Travis, and A. A. Lacis, Scattering, absorption, and emission of light by small particles. Cambridge University Press, 2002.

[23] W. Li, R. A. Jones, S. C. Allen, J. C. Heikenfeld, and A. J. Steckl, "Maximizing Alq/sub 3/OLED internal and external efficiencies: charge balanced device structure and color conversion outcoupling lenses," *Journal of Display Technology*, vol. 2, no. 2, pp. 143–152, 2006.

[24] S.-S. Sun and L. R. Dalton, Introduction to organic electronic and optoelectronic materials and devices. CRC Press, 2008.

[25] "Introduction to OLED technology."www.osram.com/media/resource/hires/335716/introduction-to-oled.

[26] J. M. M. Martin, "Charge transport in organic semiconductors with application to optoelectronic devices," PhD, Departamento de Fisica, Universitat Jaume I, Online, 2010. [Online]. Available:www.tesisenred.net/bitstream/handle/10803/10474/montero.pdf?sequence=1.

[27] A. Fluxim, "Semiconducting thin film optics simulator SETFOS," Website: www.fluxim.com.

[28] J. Ràfols-Ribé et al., "High-performance organic light-emitting diodes comprising ultrastable glass layers," *Science Advances*, vol. 4, no. 5, p. eaar8332, 2018.

[29] J. Park, Y. Kawakami, and S.-H. Park, "Numerical analysis of multilayer organic light-emitting diodes," *Journal of Lightwave Technology*, vol. 25, no. 9, pp. 2828–2836, 2007.

[30] J. Park, S. Park, and D. Shin, "Electrical properties of trilayer organic light-emitting diodes with a mixed emitting layer," *Journal of Lightwave Technology*, vol. 27, no. 13, pp. 2525–2529, 2009.

[31] L. Hung and C. Chen, "Recent progress of molecular organic electroluminescent materials and devices," *Materials Science and Engineering: R: Reports*, vol. 39, no. 5–6, pp. 143–222, 2002.

[32] C. Adachi, K. Nagai, and N. Tamoto, "Molecular design of hole transport materials for obtaining high durability in organic electroluminescent diodes," *Applied Physics Letters*, vol. 66, no. 20, pp. 2679–2681, 1995.

[33] Z. Popovic et al., "Life extension of organic LED's by doping of a hole transport layer," *Thin Solid Films*, vol. 363, no. 1–2, pp. 6–8, 2000.

[34] J. Shen et al., "Degradation mechanisms in organic light emitting diodes," *Synthetic Metals*, vol. 111, pp. 233–236, 2000.

[35] P. Burrows, V. Bulovic, S. Forrest, L. S. Sapochak, D. McCarty, and M. Thompson, "Reliability and degradation of organic light emitting devices," *Applied Physics Letters*, vol. 65, no. 23, pp. 2922–2924, 1994.

[36] M. Fujihira, L. M. Do, A. Koike, and E. M. Han, "Growth of dark spots by interdiffusion across organic layers in organic electroluminescent devices," *Applied Physics Letters*, vol. 68, no. 13, pp. 1787–1789, 1996.

[37] "Spin coat." www.brewerscience.com/processing-theories/spin-coat/ (accessed).

[38] S. S. A. Coe-Sullivan, "Hybrid organic/quantum dot thin film structures and devices," Massachusetts Institute of Technology, 2005.

[39] J. N. Bardsley, "International OLED technology roadmap: 2001–2010," US Display Consortium, 2002.

[40] A. Sharma and T. Das, "Highly efficient OLED device based on the double emissive layer with an EQE about 39%," *Optik*, vol. 221, p. 165350, 2020.

[41] A. Sharma and T. Das, "Light extraction efficiency analysis of fluorescent OLEDs device," *Optical and Quantum Electronics*, vol. 53, no. 2, pp. 1–11, 2021.

[42] A. Sharma, S. Bhattarai, and T. Das, "Efficiency improvement of organic light-emitting diodes device by attaching microlens arrays and dependencies on the aspect ratio," *Indian Journal of Physics*, pp. 1–12, 2022.

[43] B. Ruhstaller, T. Beierlein, H. Riel, S. Karg, J. C. Scott, and W. Riess, "Simulating electronic and optical processes in multilayer organic light-emitting devices," *IEEE Journal of Selected Topics in Quantum Electronics*, vol. 9, no. 3, pp. 723–731, 2003.

[44] R. Chance, A. Prock, and R. Silbey, "Molecular fluorescence and energy transfer near interfaces," *Advances in Chemical Physics*, vol. 37, pp. 1–65, 1978.

[45] A. Sharma and T. Das, "Property of fluorescent host material Alq3 organic light emitting diode device," *Adv Appl Math Sci*, vol. 18, no. 9, pp. 931–9, 2019.

[46] A. Sharma, S. Bhattarai, and T. Das, "Fluorescent trilayer OLED device: An electrical and optical characterization-based simulation," in *AIP Conference Proceedings*, 2020, vol. 2269, no. 1: AIP Publishing LLC, p. 030049.

2 Physical Mechanisms That Limit the Reliability of LEDs

Tulasi Radhika Patnala, N. Hemalatha, Sankararao Majji and M. Sundar Rajan

CONTENTS

2.1 INTRODUCTION

Semiconductor chips that have been doped with impurities are the building blocks of an LED. There is no current flow from the cathode to the anode in a diode. Electrodes with varying voltages supply the junction with charge-carriers (electrons and holes). Electrons collide with a hole, which causes them to descend into a lower energy state, releasing a photon of energy. It is the band-gap energy of the *p-n* junction materials that determines the wavelength of light emitted and thus the colour. Because silicon and germanium have indirect band gaps, the electrons and holes in these semiconductor diodes recombine in a non-radiative transition, with no visible optical emission. Depending on the material, the LED can emit light that is near-infrared, visible, or even near-ultraviolet.

Gallium arsenide-based infrared and red LEDs were the first to see widespread use. Light-emitting devices with ever-shorter wavelengths are now possible because of advances in materials science. With an electrode placed on the *p*-type layer of the *n*-type substrate, LEDs are typically made. Despite their rarity, *p*-type substrates exist. A sapphire substrate is used in many commercial LEDs, particularly in GaN/ InGaN LEDs. LED efficiency can be improved by using materials that are transparent to the emitted wavelength and are covered with a reflecting and light-spreading layer. Refractive index mismatches between the package material and semiconductor can cause light emitted to be reflected back into the semiconductor, where it can be

DOI: 10.1201/9781003340577-2

further absorbed and converted to heat, reducing the device's overall efficiency. When the LED is connected to a material with a different refractive index, such as glass fibre or air, this form of reflection also happens at the package's surface. A significantly lower refractive index medium is nearly often used because the LED's refractive index is so high. One of the most common causes of LED inefficiency is the big index difference, which results in a significant amount of reflection (based on Fresnel coefficients). The LED-package and package-air contacts often reflect more than half of the emitted light. Most frequently, the reflection is minimized by placing the diode in the centre of a dome-shaped (half-sphere) package, which directs the light rays so that they strike the surface perpendicularly. Additionally, an anti-reflective coating can be used. The colour of the container has no significant impact on the colour of the light emitted, even if it is made of low-cost plastic and coloured to improve the contrast. The LED can be designed to absorb and re-emit the reflected light (a process known as "photon recycling"), and the tiny structure of the surface can be manipulated to reduce reflectance by introducing random roughness or programming moth eye surface patterns.

When compared to other types of lighting, the flexibility of LEDs, their low power consumption, and their high level of reliability make them one of the most popular lighting options available today. Light engine and power electronic driver are two of the most common subsystems in LED luminaire lighting systems. In order to create the LED light engine, the LEDs are connected in series and parallel configurations on the printed circuit board (PCB). Additional heat created by the LEDs is dissipated by placing the PCB on a heat sink. A fibre/metal framework encloses the entire system. In order to reduce the LED light's glare or brightness, a diffuser constructed of acrylic will be employed. The LED circuitry glows at particular brightness levels thanks to the continuous current provided by the power electronic driver. For the LED luminaire, the driver comprises of a DC-DC constant current converter, an AC-DC converter, and a parallel electrolytic capacitor. LED technology has eclipsed all previous lighting technologies because of its capacity to last for 50000 hours or longer. LED luminaires' low lumen degradation and, as a result, their long lifespan have made them popular for general public lighting. While it is true that LED luminaires can last longer than single LEDs, new research reveals that the loss of LED driver electronics will result in LED luminaire failure. A decade ago, the assertion was widely accepted and now LED luminaires are mostly compared to their driver electronics in terms of life expectancy. LED luminaire driver electronic systems can now offer lifetime performance comparable to or even better than that of a single LED because of recent advances in power electronics components. According to recent studies, LED luminaire failures are primarily caused by the LEDs themselves, with driver electronics failures serving as a close second. Start-up and new LED luminaire manufacturing companies have recently taken over the lighting market, and their primary purpose is to sell their products at a reduced cost to consumers. The LEDs utilized are of poor quality, which lowers the cost and the lifetime of the overall system. As a result, research into the quality of LEDs utilized and how they work as a system is critical if we are to gain an understanding of LED luminaire lifetime performance.

2.2 LITERATURE REVIEW

Samples of Xisha Islands and a single LED outdoor lamp are used in this study to conduct research and analysis on the dependability of its use. LED light dependability under high temperature and high humidity stress was expedited prior to looking at the temperature and humidity circumstances of the Xisha Islands. In the end, the LED bulbs on the Xisha Islands were collected and analyzed, and their life expectancy was evaluated as well. Temperature and humidity variations, as well as other factors, such as salt precipitation and sun radiation, have an impact on LED lamps' durability in harsh environmental settings. It is necessary to take additional precautions when designing LED luminaires, such as taking into account the influence of salt spray, corrosive gases, and other circumstances on the service life.

According to B. Sun et al., the randomness of LED lumen degradation affects the overall LED lamp's dependability. The method's carrier is an embedded LED light bulb. For reliability prediction, a physics-of-failure (PoF)-based lumen deterioration model and electronic-thermal simulations are used. Distribution of LEDs is characterized by the normal distribution. Monte Carlo simulations can then be used to determine the likelihood of catastrophic failures and lumens by taking into account the increase in lamp temperature. Two scenarios are used to examine the impact of lumen depreciation on the lamp as a whole: constant light mode and constant current mode.

LED luminaires for high-temperature applications, such as those seen in noisy or dangerous industrial settings, are the focus of a new paper by H. Zhang and colleagues. Using an upgraded MIL-HDBK-217F method and the results of accelerated life testing, the reliability and lifespan of various LED driving technologies may be analyzed and estimated in order to optimize the design for high reliability and long-term durability. To function at greater temperatures and to have longer lifespans at the same temperature, the AC-direct driving method outperforms standard switched-mode driving for LED luminaires, as shown in the paper's comparison of the two methods for high-temperature situations. AC-direct drive circuits for high-temperature applications can also be improved to extend their service life. Finally, an AC direct driving LED luminaire with 4000 lumens and 2–3 times longer lifetime at 65°C has been developed, and it can safely operate at 80°C ambient temperature.

Effective field reliability measurement requires a combination of analytical methodologies and the ability to focus in on certain subpopulations. Digital projectors can be used for a wide variety of tasks and by a wide variety of people. As a result, the population of LED projectors is extremely heterogeneous. It's much easier to tailor customer service when you can tell the changes in performance between different types of users, clients, and even different manufacturing eras. Analytical procedures must be developed for truncation when left-censored data is used, so that relevant reliability measures may be generated and no data is left untouched. To keep track of a fleet of over 900 LED projectors and their on-site reliability, an interactive dashboard was built. The method utilized to deal with left truncation of repair data is given special attention. Parametric and non-parametric tools for repairable systems can be found on the dashboard. Age-dependent reliability performance can be analyzed with the help of the mean cumulative function (MCF).

A new generation of environmentally friendly green light sources, LEDs, are becoming more and more popular. There is a risk that cracking of the adhesive that encapsulates LEDs, particularly white LEDs that are exposed to light and heat, could develop. This will have a negative impact on light efficiency and colour temperature, potentially leading to dead lights. LED encapsulation glue cracking has been linked to two distinct causes, which are discussed in detail in this work. As a result of light and heat being absorbed by the LED lamp beads for such an extended period of time, the glue that holds the LEDs in place hardens and cracks. Another reason is that the encapsulating adhesive cracks during thermal expansion because of a thermal mismatch between the adhesive and the plastic section of the lead frame.

System integration is inextricably tied to reliability as a fundamental scientific and technological topic. Since product development and qualification cycles are getting shorter and shorter, semiconductor manufacturers have to deal with ever-greater challenges when it comes to meeting customer demands for higher levels of quality, robustness, and reliability while also having to contend with ever-decreasing design margins. Many micro-/nano-related technological developments will fail unless we innovate and make breakthroughs in the way we address reliability throughout the whole value chain. 'Design for Reliability (DfR)' is a term used to describe the process of predicting, optimizing, and designing micro-/nano-electronics and systems' reliability. However, despite their widespread use in functional design, numerical simulations lack a systematic method to reliability assessments. In addition, previous standards still assume a constant failure rate behaviour when making lifetime estimates. In this study, we'll discuss the reliability and faults that have been identified in solid-state lighting. As a result of comprehensive usage of acceleration tests and knowledge-based qualifying procedures, a complete mechanism is described, including degradation and catastrophic failure scenarios.

2.3   FAILURE MODES IN LED

Slow decline in light output and efficiency are the two most typical causes of LED (and diode) failure. However, unanticipated failures are possible. As heat, high current density, and light emitted by the crystal all work together, they help to accelerate processes like nucleation and dislocation formation in the active zone. This is where radiative recombination takes place after all. It's necessary for this method to work if gallium and aluminum are less susceptible than gallium arsenide phosphide and indium arsenide. It is also possible for ionizing radiation to cause such faults, which can lead to radiation hardening concerns in LED circuits (e.g., in opto-isolators). The limited life expectancy of early red LEDs attracted attention.

One or more phosphors may be used in the production of white LEDs. Heat and time destroy the phosphors, reducing their effectiveness and altering the hue of the light they generate. Organic phosphor formulations used in pink LEDs may decay after just a few hours of operation, resulting in a significant hue shift. Metal atoms can diffuse from the electrodes into the active zone when subjected to high electrical currents and high temperatures. As a result, leakage current and non-radiative recombination occurs at the chip edges due to electro-migration of certain materials.

An electromigration barrier metal layer may be utilized in some GaN/InGaN diodes to prevent degradation of the device performance. Short circuits can be caused by whiskers forming due to mechanical stress, high currents, and a corrosive environment.

Current crowding, or an uneven distribution of current density over the junction, is a problem with high-power LEDs. Thermal runaway can occur as a result of the development of localized hot spots. Localized loss of thermal conductivity due to non-homogeneities in the substrate is made worse by voids created by improper soldering, electro-migration effects, and Kirkendall voiding, which are the most common causes of this problem. LED failures are frequently caused by thermal runaway. To avoid catastrophic optical damage, the light output of laser diodes must be kept under control and the facet melted. It's possible that some of the plastic package's materials will turn yellow when heated, resulting in a reduction of efficiency in the wavelengths impacted. Thermal stresses are the most common cause of unexpected breakdowns. A semiconductor and its bonded contact can be weakened or torn apart when epoxy resin used in packaging is heated to its glass transition temperature and then allowed to rapidly expand. The packaging can shatter if it is exposed to extremely low temperatures. Electrostatic discharge (ESD) may induce a temporary shift in the characteristics of the semiconductor junction or latent damage that increases the deterioration rate. These devices are more vulnerable to ESD damage when fabricated on sapphire substrates.

2.4 LIGHT-EMITTING DIODES (LEDS) ON THE MARKETS

There are many components to high-power LEDs, including a blue GaN-based LED chip with an emission wavelength of 450–460 nm at its core. In order to effectively dissipate heat, this chip, which is typically 1 mm^2 in size, is housed in a power LED package atop a thermally conductive frame. A lens is typically placed over the power LED chip to improve light extraction efficiency while also altering the shape of the light beam emitted. Chip-level conversion (CLC) or incorporation within the lens are both options for converting blue light to white light via a phosphorous layer.

These commercially available chips typically operate at a current of 350–1000 mA, with a voltage of 3.2–3.4 V. These devices, then, consume between 1 and 3.5 watts of electricity. When used in their final application, power LEDs have a thermal resistance of 10–20 K/W, depending on how well the heat extraction path is optimized. Because of this, devices can demonstrate a degree of self-heating, with a temperature increase of between 20°C and 80°C when functioning in typical settings. The current and temperature levels LEDs experience as they age can have a significant impact on their deterioration kinetics: As a result, attaining a long LED lifespan necessitates meticulous heat dissipation process optimization and correct characterization of operating parameters.

Experiments in which LEDs are treated to relatively high current and/or temperature settings in order to expedite their ageing are rather common. This makes it difficult to pinpoint the root cause of degradation because of the high current density flowing through and the high operating temperatures of the electronics involved here. However, even if the LED structure is degraded simultaneously with the blue LED chip's ageing, it is impossible to tell whether the degradation is due to the processes associated to its package/phosphors/lens system. In order to develop particular testing methodologies for evaluating the degradation of LED components over time, physical reasons that limit LED structural dependability must be properly studied. Below you'll find basic findings about the dependability of modern white LEDs and statistics on the degradation of blue LED chips and packages/phosphors systems. White LEDs have been demonstrated to be dependable in several applications.

2.5 HIGH-POWER WHITE LEDS ARE SUBJECTED TO RELIABILITY TESTING

A somewhat accelerated environment is typically used to conduct ageing studies on high-power LEDs. It is possible to do this by ageing the LEDs at the maximum current given in the datasheets at room temperature. The junction-to-ambient thermal resistance is kept as low as feasible during stress tests by attaching devices to heat sinks with thermally conductive adhesives. E. F. Schubert's research group created an indirect forward voltage method that can be used to calculate the thermal resistance of LEDs. Except for the time required for measurement, the ageing current should be delivered constantly. When the luminous flux drops by 10%–30% or the operating voltage shifts by 200 mV or more, the device is deemed faulty. A benchmarking test at 100 A/cm2, 80 °C yielded the following deterioration curves for three generations of cutting-edge white LEDs. The mid-term stress test showed that one of the three LEDs under evaluation had a considerable loss in optical power, which suggests a high susceptibility to stress. The reliability of various device technologies can be compared using the results of ageing experiments like this one or comparable ones. On the other hand, monitoring optical power depreciation (or operational voltage change) during stress tests will yield no substantial information on the physical process responsible for degradation. Specific stress testing on appropriate test and device constructions are required to determine the cause of lumen depreciation. To better understand the degradation mechanisms that influence the LED chip, the package, and the phosphors of modern LED structures, we conducted a series of stress tests to examine the effects of these factors on LED efficiency and optical qualities. Figure 2.1 shows the degradation of different generations of LEDs with stress time.

2.6 INGAN LED CHIP DEGRADATION

LED chips based on InGaN were subjected to constant current stress in order to characterize the deterioration of the LED chips available commercially and in research. Peltier-controlled fixtures were used to accurately control the temperature of the case during the stress tests. Case temperatures varied from 15 to 85°C, with a stress current ranging from 20 to 180 A/cm^2. In order to understand how the devices degraded, a thorough optical and electrical characterization was performed throughout the stress tests. The samples' optical power can be reduced when they are under stress, as can be seen. Low measurement current has a greater impact, as shown in Figure 2.2's optical power/stress time curves. The active layer's non-radiative recombination centres can lower the quantity of electrons that undergo radiative recombination by limiting the

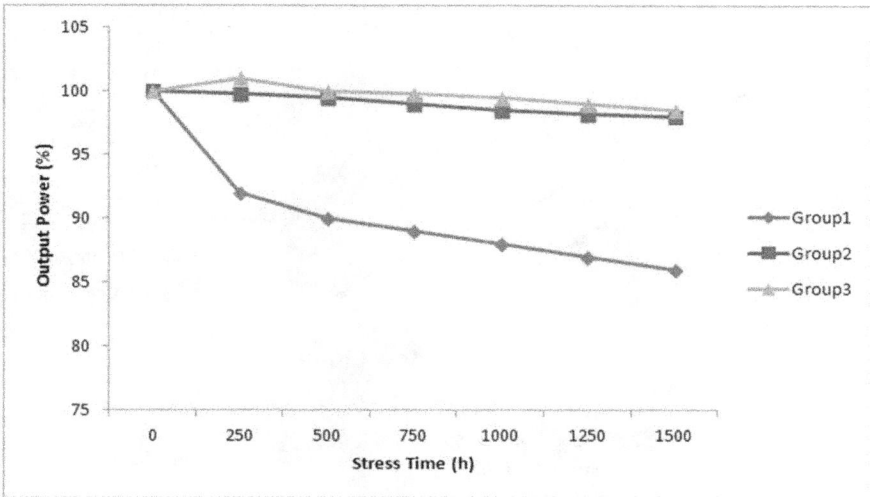

FIGURE 2.1 Three generations of white LEDs' optical power degradation curves.

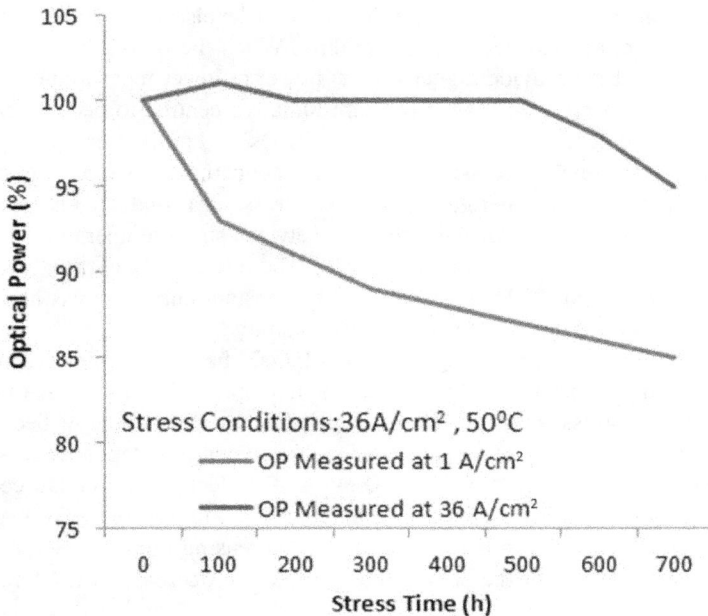

FIGURE 2.2 One of the bare InGaN-based LEDs examined in this study had its optical power degradation assessed under stress at a constant current level.

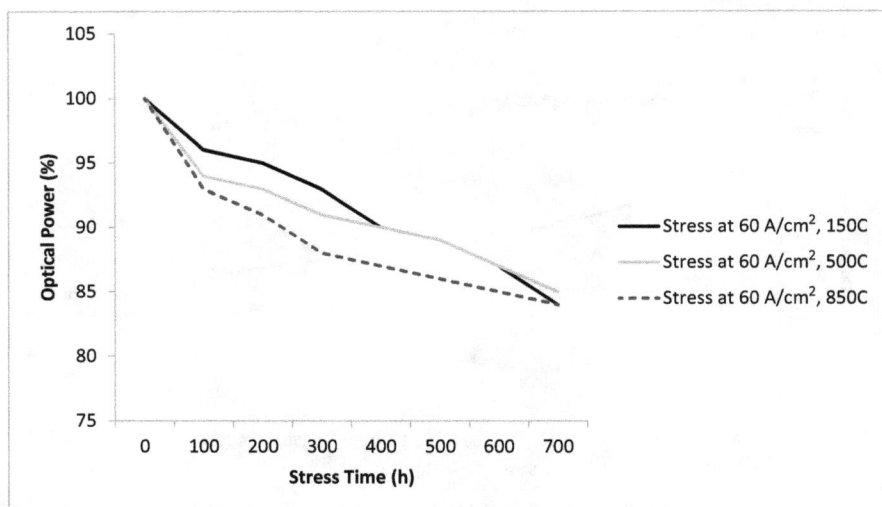

FIGURE 2.3   Stress-induced power loss in varied instance temperatures.

flow of electrons across the junction at low current levels. Stress has increased the non-radiative rate, as evidenced by this finding. When the non-radiative recombination rate rises, LEDs provide higher optical power at lower measurement currents. A large quantity of electrons can cause non-radiative centres to become saturated; hence high current levels have no influence on LEDs' *L-J* properties.

As far as commercially accessible samples are concerned, this type of degradation process is only moderately thermally activated. Stress tests conducted at various case temperatures showed only a weak correlation between stress temperature and deterioration kinetics (see Figure 2.3 for an example). There is a strong correlation between the degradation of bare LED chips exposed to constant current stress and current, rather than temperature, as demonstrated by this study.

Recent findings describe the degradation of InGaN-based laser diodes, both in the lasing and subthreshold operating ranges, which supports this notion. A linear relationship between stress current level and deterioration rate was established for these devices; on the other hand, degradation had weak temperature dependence with activation energies in the range of 0.25–0.30 eV. As the electrons in the lattice interact with the highly energetic electrons, defects are commonly created or propagated as a result of this interaction in the same manner. Increasing non-radiative recombination in the active layer of the devices could lead to a decrease in light generation efficiency.

Furthermore, capacitance-voltage profiling has been used to show that the charge distribution in the active zone can radically change under continual current stress. *C-V* measurements can be used to analyze the distribution of charged carriers in the active layer of LEDs. These data allow researchers to infer the junction's apparent charge distribution (ACD) and look for stress-induced variations. Stress resulted in a considerable rise in junction capacitance between–4 and 0 V, as can be shown. The

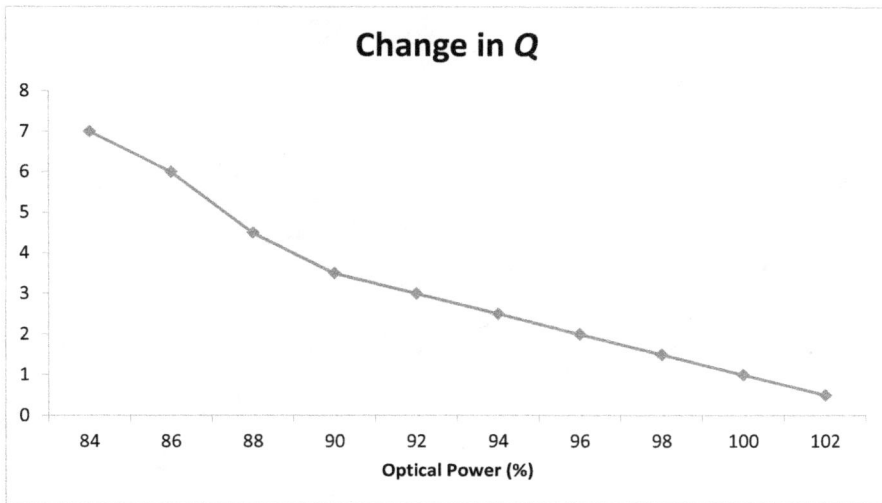

FIGURE 2.4 Charge concentration in the quantum well area versus optical power loss on an InGaN-based LED under constant current stress.

quantum-well area is where the space charge border is placed when the voltage is set in this range. When LEDs are operated at this voltage, the active layer charge densities move significantly, which is consistent with an increase in junction capacitance. The amount of charge that the continuous current stress experiment supplied can be estimated by integrating the rise in differential capacitance. Figure 2.4 depicts the most significant outcome of this investigation. When a constant current stress induces an increase in optical power and decrease in charge concentration, this graph shows the resulting changes on the horizontal axis.

Researchers have found that InGaN-based LED chips also degrade when reverse-biased. Stress tests were conducted on a number of InGaN-based LEDs to investigate the deterioration mechanisms generated by reverse-biasing. Tests were conducted using stress voltage levels between–20 V and–45 V as well as a range in stress current density from–0.3 A/cm² to–2.4 A/cm².

In the quantum-well region, optical power loss and changes in charge concentration have a nearly linear relationship, according to the results of this study. Because of this, the amount of defects in the LEDs' active zone is exactly proportional to the rate of non-radiative recombination, making it important to keep this in mind while interpreting the results. As a result, by increasing the density of charged defects, it is possible to reduce optical efficiency while simultaneously increasing the apparent charge concentration in quantum-well regions. Deep levels with activation energies ranging from 160 to 430 meV have been connected to these alterations in charge distribution. LED degradation cannot be reliably traced back to a single cause, but a number of studies are underway to try. Researchers using secondary ion mass spectroscopy (SIMS) believe that LED deterioration may be caused by dopant penetration into the active layer of LEDs activated by stress tests. Dislocations in the

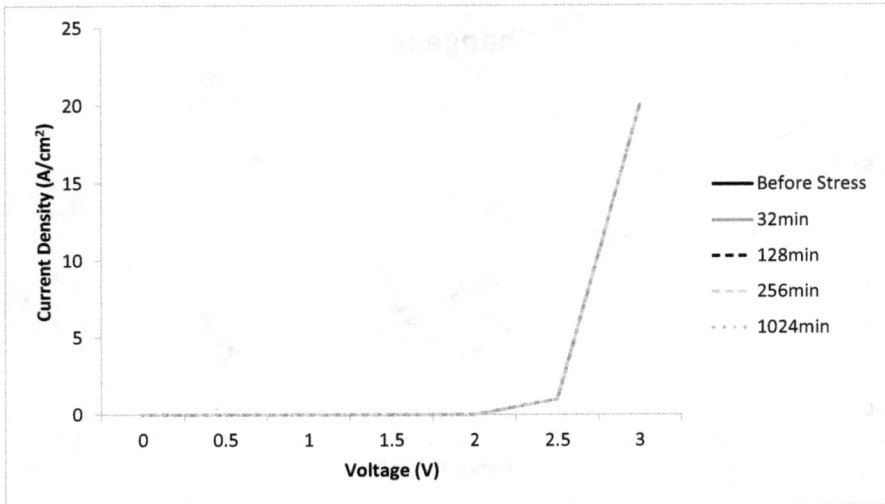

FIGURE 2.5   Forward-bias current density versus voltage characteristics of a negative-biased InGaN-based LED.

quantum well region, in particular, could lead to a considerable rise in non-radiative components and a decrease in device efficiency, according to researchers.

Researchers have found that InGaN-based LED chips also degrade when reverse-biased. Stress tests were conducted on a number of InGaN-based LEDs to investigate the deterioration mechanisms generated by reverse-biasing. Tests were conducted using stress voltage levels between–20 V and–45 V as well as a range in stress current density from–0.3 A/cm^2 to–2.4 A/cm^2.

When LEDs are subjected to a negative bias, they display a significant rise in reverse current. An increase in reverse current indicates that the breakdown voltage has decreased significantly. The electrical characteristics of InGaN-based LEDs (as well as their forward-bias operating voltage) are unaffected by reverse-bias stress, on the other hand (Figure 2.5).

As a result of reverse-bias stress, current flows through localized leakage channels, which are linked to the occurrence of faults in LEDs. Under reverse-bias, the high electric field accelerates injected carriers substantially, allowing them to interact with the lattice and generate point defects. Leakage routes responsible for reverse current conduction are where defects are generated. LEDs subjected to negative-bias stress experience a large rise in reverse leakage current as a result of this mechanism. As a result of reverse-bias stress, LED forward-bias electrical and optical characteristics have not changed much. Considering that (i) deterioration affects nanometer-sized sections, which are significantly smaller than the entire device area; and (ii) under forward bias, a junction's impedance is lower than the impedance of newly formed defective leakage channels. This means that LED forward bias behaviour is dominated by the LED junction's conduction and recombination mechanisms, rather than the existence of defective tunnelling channels.

2.7   DIFFUSION-RELATED DEGRADATION PROCESSES IN LED

The lifespan of modern products has increased due to advancements in material science and manufacturing technologies, making it difficult to gather adequate failure data during the research and development phase. Traditional reliability analysis methods have been unable to adequately analyze the reliability of these goods. Reliability analysis can make use of deterioration data in this case. They are easier to collect than failure data. There has been an explosion in the use of deterioration data for dependability assessment in the previous few decades.

For many electrical and electromechanical products, degradation is a typical occurrence that may be defined as a continuous performance process in terms of time. Linear regression, degradation path, and stochastic process approaches have been proposed in earlier research for the study of degradation data. Stochastic processes, such as the Gamma process, Wiener process, and Markov process, have been frequently utilized to predict deterioration paths because they can explain failure mechanisms and operating environment features in a flexible fashion.

Tseng et al. used the Wiener process to estimate LED lamp lifetimes and Peng and Tseng used the Wiener process to simulate GaAs lasers' reliability assessments in recent years. The Wiener process can only explain a linearly drifting diffusion process, which is a basic difficulty. However, in practice, non-linearity is widespread, and the linear model is unable to capture the dynamics of the deterioration of non-linearity. As a result, non-linear structures are necessary in a realistic model.

The log-transformation or time-scale transformation of deterioration data can mimic some non-linear processes linearly. It's possible to transform some non-linear deterioration processes, but not all of them can. Because each product may have different causes of deterioration, random effect models are more appropriate to capture unit-to-unit variability. Only the non-linear processes that could be turned into linear processes by the time-scale transformation were tested in Wang's article, which focused on Wang's random-effect Wiener process model. In order to better capture the degradation process, it may be preferable to use a random effect non-linear diffusion model without the aforementioned adjustments.

As a result of the increased functionality and complexity of modern products, several degradation mechanisms may be at play. For example, the LED system may generate several deterioration mechanisms that lead to failure when used for distinct purposes of lighting. Only a little amount of work has been done on multivariate PCs. In these studies, the numerous PCs are assumed to be independent, or the multivariate normal distribution is used. These assumptions may not be reasonable from a real-world perspective. A copula function has recently been utilized to reduce these assumptions and express the reliance of PCs. According to Sari et al., how can a system with two or more deteriorated PCs be quantified in terms of its reliability? These PCs can be characterized by a copula function.

On the other hand, Sari's study utilized a generalized linear regression model based on the population average technique. Stochastic process models have a wider range of applications than regression models due to their ability to account for temporal uncertainty in a degradation process. We used the Wiener process and the copula function to simulate degradation failure mechanisms and to characterize the dependency on

Sari's work for different degradation failure mechanisms. They followed Sari's lead and used a population average technique, but they failed to take the random effect into account.

A wide range of applications, including UV curing, phototherapy, and plant growth illumination, can benefit from UV LEDs based on Al(In)GaN with wavelengths between 280 and 320 nm. UVB lights have a much shorter lifespan than other LEDs, even when compared to UVA and blue LEDs. An important first step in resolving this problem is to gain an understanding of the numerous degradation processes that occur within LEDs when they are operating. Temperature and current may affect these processes in different ways, allowing for their differentiation.

Constant current operation for UVB LEDs has already been proven to lower the optical power and that larger current densities lead to a faster loss of optical power for UVB LEDs. As with nitride-based visible light emitting diodes (LEDs), a current-driven deterioration was identified, however there are studies that show that the impact of current on LED performance is negligible when compared to the temperature. An exponent of 0.17 has been discovered to have a significant effect on LEDs' lifetimes, such as white LEDs.

Using typical chip processing techniques, the LEDs were created after the epitaxial growth process. When it comes to electrode materials, platinum is used for the p electrodes, while vanadium and aluminum are used for the n electrodes. Flip-chip LED chips were put on AlN ceramic packages using AuSn solder for enhanced heat extraction. In other words, no encapsulation was used, and the package was left open. A PCB with an aluminum core was used to solder the LEDs, and screws were used to attach the PCB to the heat sink. Thermal resistance of roughly 24 K/W was discovered between the pn connector and the heat sink.

The LEDs emit a single peak of light at a wavelength of 350 nm. This difference is roughly two orders of magnitude greater before and after ageing in the QW emission peak intensity. As a result, the QW emission can be attributed to the measured relative optical power during operation. Parasitic emission levels vary from sample to sample, but there is no consistent pattern that can be tied to the conditions under which they were collected. Stress-induced variations in the wavelength or half-width of the emission peak are either non-existent or too tiny to be spotted.

The optical power decreases more rapidly at the beginning of operation than it does during operation, as has been shown in prior studies. This holds true for all of the study's current densities. It's interesting to note that at high current densities, degradation rates are nearly independent of time.

2.8 CONCLUSION

In this chapter, we examined the physical processes that restrict the resiliency of GaN-based LEDs and found some interesting results. LEDs of the latest generation are multicomponent devices. Each one of these components can be damaged by stress. LEDs, for example, can lose light flux and develop chromatic aberrations as they age. LEDs' active layer current flow may boost non-radiative recombination. It is possible to link LED degradation to a significant adjustment in the apparent charge distribution in the quantum-well area, which may be created by a rise in deep-level concentration or the diffusion of dopant or impurities into the active layer of the LEDs' active layer.

UV LEDs emitting light at 310 nm have a time-dependent optical power dependence that may be modelled using the operational current density cube. The higher the current density, the shorter the life expectancy of these gadgets. By reducing the current density, for example, the longevity can be increased by eight times. When it comes to improving reliability, a larger (p-contact) area can help. Auger recombination was also implicated in the degradation, according to certain theories. Increasing the amount of QWs or adjusting the carrier distribution in the active region, for example, can help increase the lifetime.

REFERENCES

Apostolou, G., A. Reinders and M. Verwaal, "Comparison of the Indoor Performance of 12 Commercial PV Products by a Simple Model". Energy Sci. Eng., vol. 4, pp. 69–85, 2016. doi: 10.1002/ese3.110.

Bobashev, G., N. G. Baldasaro, K. C. Mills and J. L. Davis, "An Efficiency-Decay Model for Lumen Maintenance", IEEE Trans. Device Mater. Rel., vol. 16, no. 3, pp. 277–281, Sep. 2016.

Bothe, K. M., et al., "Improved X-Band Performance and Reliability of a GaN HEMT With Sunken Source Connected Field Plate Design", in IEEE Electron Device Lett., vol. 43, no. 3, pp. 354–357, March 2022, doi: 10.1109/LED.2022.3146194.

Cai, M., et al., "Junction Temperature Prediction for LED Luminaires Based on a Subsystem-Separated Thermal Modeling Method", IEEE Access, vol. 7, pp. 119755–119764, 2019.

Deshayes, Y., L. Bechou, F. Verdier and Y. Danto, "Long-Term Reliability Prediction of 935 nm LEDs using Failure Laws and Low Acceleration Factor Ageing Tests", Qual. Rel. Eng. Int., vol. 21, no. 6, pp. 571–594, Oct. 2005.

Elger, G., and M. Schmid, "Reliability of SAC+ Solders for LED Packages", 2021 27th International Workshop on Thermal Investigations of ICs and Systems (THERMINIC), 2021, pp. 1–6, doi: 10.1109/THERMINIC52472.2021.9626502.

Enayati, J., and Z. Moravej, "Real-Time Harmonic Estimation using a Novel Hybrid Technique for Embedded System Implementation", Int. Trans. Electr. Energy Syst., vol. 27, no. 12, Dec. 2017.

Enayati, J., and Z. Moravej, "Real-Time Harmonics Estimation in Power Systems using a Novel Hybrid Algorithm", IET Gener. Transm. Distrib., vol. 11, no. 14, pp. 3532–3538, Sep. 2017.

Enayati, J., A. Rahimnejad and S. A. Gadsden, "LED Reliability Assessment Using a Novel Monte Carlo-Based Algorithm", in IEEE Trans. Device Mater. Reliab., vol. 21, no. 3, pp. 338–347, Sept. 2021, doi: 10.1109/TDMR.2021.3095244.

Fan, X., W. Guo and J. Sun, "Reliability of High-Voltage GaN-Based Light-Emitting Diodes," in IEEE Trans. Device Mater. Reliab., vol. 19, no. 2, pp. 402–408, June 2019, doi: 10.1109/TDMR.2019.2917005.

Fan, J., K.-C. Yung and M. Pecht, "Predicting Long-Term Lumen Maintenance Life of LED Light Sources using a Particle Filter-Based Prognostic Approach", Expert Syst. Appl., vol. 42, no. 5, pp. 2411–2420, 2015, [online] Available: https://doi.org/10.1016/j.eswa.2014.10.021.

Fan, J., K.-C. Yung and M. Pecht, "Prognostics of Lumen Maintenance for High Power White Light Emitting Diodes using a Nonlinear Filter-Based Approach", Rel. Eng. Syst. Safety, vol. 123, pp. 63–72, Mar. 2014, [online] Available: https://doi.org/10.1016/j.ress.2013.10.005.

Fettke, M., et al., "A Study on Laser-Assisted Bonding (LAB) and its Influence on Luminescence Characteristics of Blue and YAG Phosphor Encapsulated InGaN LEDs", 2020 IEEE 70th Electronic Components and Technology Conference (ECTC), 2020, pp. 1928–1934, doi: 10.1109/ECTC32862.2020.00301.

Foti, M., C. Tringali, A. Battaglia, N. Sparta, S. Lombardo and C. Gerardi, "Efficient Flexible Thin Film Silicon Module on Plastics for Indoor Energy Harvesting". Sol. Energy Mater. Sol. Cells., vol. 130 pp. 490–494, 2014. Doi: 10.1016/j.solmat.2014.07.048.

Fu, H.-K., S.-P. Ying, T.-Te Chen, H.-H. Hsieh and Y.-C. Yang, "Accelerated Life Testing and Fault Analysis of High-Power LED", IEEE Trans. Electron Devices, vol. 65, no. 3, pp. 1036–1042, Mar. 2018.

Girish, T., "Some Suggestions for Photovoltaic Power Generation using Artificial Light Illumination". Sol. Energy Mater. Sol. Cells., vol. 90, pp. 2569–2571, 2006. Doi: 10.1016/j.solmat.2006.03.026.

Goebel, K., B. Saha and A. Saxena, "A Comparison of Three Data-Driven Techniques for Prognostics", Proc. 62nd Meeting Soc. Mach. Failure Prevent. Technol. (MFPT), pp. 119–131, 2008.

Gorcea, E., and M. Wieder, "LED Projector Field Reliability Evaluation Tool Using Left-Censored Data", 2019 Annual Reliability and Maintainability Symposium (RAMS),2019, pp. 1–7, doi: 10.1109/RAMS.2019.8769285.

Hao, J., Q. Sun, Z. Xu, L. Jing, Y. Wang and H. L. Ke, "The Design of Two-Step-Down Aging Test for LED Lamps under Temperature Stress", IEEE Trans. Electron Devices, vol. 63, no. 3, pp. 1148–1153, Mar. 2016.

Ibrahim, M. S., J. Fan, W. K. C. Yung, Z. Wu and B. Sun, "Lumen Degradation Lifetime Prediction for High-Power White LEDs Based on the Gamma Process Model", IEEE Photon. J., vol. 11, no. 6, Dec. 2019.

IES Approved Method: Measuring Luminous Flux and Color Maintenance of LED Packages Arrays and Modules, New York, NY, USA, 2015, [online] Available: www.techstreet.com/standards/ies-lm–80–15?product_id=1900618.

Jarosz, G., R. Marczyński and R. Signerski, "Effect of Band Gap on Power Conversion Efficiency of Single-Junction Semiconductor Photovoltaic Cells under White Light Phosphor-Based LED Illumination". Mater. Sci. Semicond. Process., vol. 107 p 104812, 2020. doi: 10.1016/j.mssp.2019.104812.

Lall, P., and J. Wei, "Prediction of L70 Life and Assessment of Color Shift for Solid-State Lighting using Kalman Filter and Extended Kalman Filter-Based Models", IEEE Trans. Device Mater. Relab., vol. 15, no. 1, pp. 54–68, Mar. 2015.

Liu, C.-H., H.-C. Chiu, H.-C. Wang, H.-L. Kao and C.-R. Huang, "Improved Gate Reliability Normally-Off p-GaN/AlN/AlGaN/GaN HEMT With AlGaN Cap-Layer", in IEEE Electron Device Lett., vol. 42, no. 10, pp. 1432–1435, Oct. 2021, doi: 10.1109/LED.2021.3109054.

Minnaert, B., and P. Veelaert, "Efficiency Simulations of Thin Films Chalcogenide Photovoltaic Cells for Different Indoor Lighting Con-ditions". Thin Solid Films., vol. 519, pp. 7537–7540, 2011. doi: 10.1016/j.tsf.2011.01.362.

Padmasali, A. N., and S. G. Kini, "Accelerated Testing Based Lifetime Performance Evaluation of LEDs in LED Luminaire Systems", in IEEE Access, vol. 9, pp. 137140–137147, 2021, doi: 10.1109/ACCESS.2021.3118106.

Padmasali, A. N., and S. G. Kini, "A Generalized Methodology for Predicting the Lifetime Performance of LED Luminaire", IEEE Trans. Electron Devices, vol. 67, no. 7, pp. 2831–2836, Jul. 2020.

Padmasali, A. N., and S. G. Kini, "A Lifetime Performance Analysis of LED Luminaires under Real-Operation Profiles", IEEE Trans. Electron Devices, vol. 67, no. 1, pp. 146–153, Jan. 2020.

Padmasali, A. N., and S. Kini, "Prognostic Algorithms for L70 Life Prediction of Solid State Lighting", Light. Res. Technol., vol. 48, no. 5, pp. 608–623, Aug. 2016.

Pei, N., Q. Chen, R. Hu and X. Luo, "Effect of LED Chip Displacement on its Optical Performance and Reliability", 2018 19th International Conference on Electronic Packaging Technology (ICEPT), 2018, pp. 1433–1436, doi: 10.1109/ICEPT.2018.8480457.

Qian, C., J. J. Fan, X. J. Fan, A. E. Chernyakov and G. Q. Zhang, "Lumen and Chromaticity Maintenance Lifetime Prediction for LED Lamps using a Spectral Power Distribution Method", Proc. 12th China Int. Forum Solid State Light. (SSLCHINA), pp. 67–70, Dec. 2015.

Qian, C., J. Fan, X. Fan and G. Zhang, "Prediction of Lumen Depreciation and Color Shift for Phosphor-Converted White Light-Emitting Diodes Based on a Spectral Power Distribution Analysis Method", IEEE Access, vol. 5, pp. 24054–24061, 2017.

Shaygi, M., M. Li, K.-J. Lang, H. Laux and B. Wunderle, "A Numerical and Experimental Investigation of Influential Factors for Solder Joint Reliability of Power LEDs for Automotive Applications", 2022 23rd International Conference on Thermal, Mechanical and Multi-Physics Simulation and Experiments in Microelectronics and Microsystems (EuroSimE), 2022, pp. 1–9, doi: 10.1109/EuroSimE54907.2022.9758848.

Si, X.-S., W. Wang, C.-H. Hu and D.-H. Zhou, "Remaining Useful Life Estimation—A Review on the Statistical Data Driven Approaches", Eur. J. Oper. Res., vol. 213, no. 1, pp. 1–14, 2011, [online] Available: https://doi.org/10.1016/j.ejor.2010.11.018.

Singh, P., C. M. Tan, W. Zhao and H. C. Kuo, "Investigation of the Impact of Drive Current and Phosphor Thickness on the Reliability of High Power White LED Lamp", in IEEE Trans. Device and Materials Reliability, vol. 19, no. 2, pp. 290–297, June 2019, doi: 10.1109/TDMR.2019.2904740.

Soltani, M., M. Freyburger, R. Kulkarni, R. Mohr, T. Groezinger and A. Zimmermann, "Reliability Study and Thermal Performance of LEDs on Molded Interconnect Devices (MID) and PCB", in IEEE Access, vol. 6, pp. 51669–51679, 2018, doi: 10.1109/ACCESS.2018.2869017.

Song, B. M., and B. Han, "Analytical/Experimental Hybrid Approach Based on Spectral Power Distribution for Quantitative Degradation Analysis of Phosphor Converted LED", IEEE Trans. Device Mater. Rel., vol. 14, no. 1, pp. 365–374, Mar. 2014.

Sun, B., J. Fan, X. Fan and G. Zhang, "A Probabilistic Physics-of-Failure Reliability Assessment Approach for Integrated LED Lamps", 2018 19th International Conference on Thermal, Mechanical and Multi-Physics Simulation and Experiments in Microelectronics and Microsystems (EuroSimE), 2018, pp. 1–5, doi: 10.1109/EuroSimE.2018.8369897.

Sun, B., X. Jiang, K.-C. Yung, J. Fan and M. G. Pecht, "A Review of Prognostic Techniques for High-Power White LEDs", IEEE Trans. Power Electron., vol. 32, no. 8, pp. 6338–6362, Aug. 2017.

Trivellin, N., M. Meneghini, M. Buffolo, G. Meneghesso and E. Zanoni, "Failures of LEDs in Real-World Applications: A Review", in IEEE Transactions on Device and Materials Reliability, vol. 18, no. 3, pp. 391–396, Sept. 2018, doi: 10.1109/TDMR.2018.2852000.

Truong, M. T., L. Mendizabal, P. Do and B. Iung, "A Novel Degradation Model for LED Reliability Assessment with Accelerated Stress and Self-Heating Consideration", 2021

IEEE 71st Electronic Components and Technology Conference (ECTC), 2021, pp. 2258–2265, doi: 10.1109/ECTC32696.2021.00354.

van Driel, W. D., M. Schuld, B. Jacobs, F. Commissaris, J. van der Eyden and B. Hamon, "Lumen Maintenance Predictions for LED Packages", Microelectron. Rel., vol. 62, pp. 39–44, Jul. 2016, [online] Available: https://doi.org/10.1016/j.microrel.2016.03.018.

Xu, H., Y. Tang, J. Wu, B. Peng, Z. Chen and Z. Liu, "The Study on Cracking Reasons of LED Encapsulation Silicone", 2019 20th International Conference on Electronic Packaging Technology(ICEPT), 2019, pp. 1–3, doi: 10.1109/ICEPT47577.2019.245288.

Yang, N. G. M., B. Y. R. Shieh, T. F. Y. Zeng and S. W. Ricky Lee, "Analysis of Pulse-Driven LED Junction Temperature and its Reliability", 2018 15th China International Forum on Solid State Lighting: International Forum on Wide Bandgap Semiconductors China (SSLChina: IFWS), 2018, pp. 1–3, doi: 10.1109/IFWS.2018.8587392.

Yaoyang, S., and S. Bo, "The Reliability Assessment of Pulse-Driven Light Emitting Diodes", 2021 22nd International Conference on Electronic Packaging Technology (ICEPT), 2021, pp. 1–5, doi: 10.1109/ICEPT52650.2021.9568030.

Zhang, H., "Developing highly reliable LED luminaires for high temperature applications using AC-direct driving LED technology", 2018 IEEE Applied Power Electronics Conference and Exposition (APEC), 2018, pp. 3466–3470, doi: 10.1109/APEC.2018.8341602.

Zhang, H., et al., "Improved Reliability of AlGaN-Based Deep Ultraviolet LED With Modified Reflective N-Type Electrode", in IEEE Electron Device Lett., vol. 42, no. 7, pp. 978–981, July 2021, doi: 10.1109/LED.2021.3081576.

Zhang, J., et al., "Life Prediction for White OLED Based on LSM under Lognormal Distribution", Solid. State. Electron., vol. 75, pp. 102–106, Sep. 2012 [online] Available: https://doi.org/10.1016/j.sse.2011.12.004.

Zhonghong, S., W. Chunxia, D. Xiaoqun, H. Jianyao, X. Huawei and X. Jiang, "Research on Reliability of LED Luminaires in Extreme Natural Environment", 2018 19th International Conference on Electronic Packaging Technology (ICEPT), 2018, pp. 31–35, doi: 10.1109/ICEPT.2018.8480426.

3 Scattering Effects on the Optical Performance of LEDs

Vinodhini Subramaniyam, B. A. Saravanan and Moorthi Pichumani

CONTENTS

3.1   INTRODUCTION

LEDs are one of the solid-state (Li et al. 2022a) and eco-friendly lighting sources (Ryu and Shim 2010). The compact size, flexibility and cost-effective nature (Damulira et al. 2020) aid them to find applications in various fields, such as displays, traffic

DOI: 10.1201/9781003340577-3

signals, solar cells, etc., LEDs are stunning sources that could replace conventional lamps with low power consumption (Liu et al. 2010). The optical performance of LEDs is improvised by changing the shape from cubic into stripes. The stripe shape improvises the heat dissipation, followed by the improvement in the optical emission (Jin et al. 2021). When considering the optical performance, external quantum efficiency is the crucial factor to compute. The external quantum efficiency $\left(\eta_{external}\right)$ is describes the ratio between the total number of output photons and the total number of injected carriers and it is simply formulated as follows:

$$External\,Quantum\,Efficiency\left(\eta_{external}\right) = \frac{Total\,No.of\,Output\,Photons}{Total\,No.of\,Injected\,Carrier} \quad (3.1)$$

Internal quantum efficiency $\left(\eta_{internal}\right)$ is the ratio between the total number of photons produced inside the emitter and the total number of carriers (electrons) injected and it is formulated as follows:

$$Internal\,Quantum\,Efficiency\left(\eta_{internal}\right) = \frac{Total\,No.of\,Photons\,Produced\,in\,emitter}{Total\,No.of\,Injected\,Carrier}$$

$$(3.2)$$

From the equations (3.1) and (3.2),

$$\eta_{external} = \eta_{internal} \times \eta_{outcoupling} \quad (3.3)$$

where $\left(\eta_{outcoupling}\right)$ denotes the external coupling efficiency, described as the ratio between the total number of coupled out photons from the device and the total number of photons produced (Gomard et al. 2016).

From equation (3.3), the external quantum efficiency of the LED is directly related to the external quantum efficiency. Also, this external quantum efficiency is affected by several aspects, such as increased dislocation concentration, minimal hole concentration, less light/optical extraction efficiency, higher polarization effect, less conductivity, heating effect, etc. (Kumar et al. 2021). For example, in GaN-based LEDs, the optical efficiency loss is instigated by the indirect auger recombination (Kioupakis et al. 2011). Light extraction efficiency improvement is a very common problem in LEDs, due to the total internal reflection and scattering effects of the optical spectrum caused by the device structure (Zhang et al. 2014b). Importantly, this can be resolved by modifying the structure/shape of the active layer and/or the top surface of the LED device. Otherwise, it can be resolved by moving the optical emission region into the optical cavity. These actions can produce ways of utilizing the scattering effects, in order to enhance the optical performance of the LEDs (Zhmakin 2011). Scattering is one of the most important phenomena of light, which cannot be eliminated while discussing LEDs, because of the broader bandwidth (Hart and JiJi 2002). Scattering effects help in recycling the entrapped photons in the absorbent surfaces of the LED components (Cho et al. 2020).

3.2 FACTORS AFFECTING THE SCATTERING

Scattering of the photons in LEDs generally decreases the performance. However, to utilize the scattering effect positively, several methods are used to achieve the enhanced external efficiency, especially in white LEDs (Gather and Reineke 2015). This can be attained by means of several factors, mainly, scatterer-free methods and scatterer-infused methods. However, in earlier times, researchers used the scattering medium in LEDs used in *invivo* spectroscopy, which are the aqueous medium (biological medium) filled with homogeneously distributed scatterers, for example, intravenous fat emulsion (liposyn III 20%) (Fantini et al. 1994). Then, liquid-like amorphous photonic crystals are utilized as scattering medium to enhance the photon extraction efficiency (Yue et al. 2016). Afterwards, the solid films are used as scattering medium in LEDs (Pyo et al. 2016). Specifically, one or a combination of effects are assisted towards the enhancement of internal and external output efficiency in LED devices (Taniyasu et al. 2006). By using simulation methods, researchers found that the carrier concentration is the key factor, which is causing several scattering effects (Shedbalkar et al. 2016). Because, the emission in semiconductors occur between the conduction band and the valence band (Mukherjee et al. 2016). This considers the presence of the concentration of electrons and holes (Shedbalkar and Witzigmann 2018) in the semiconductor materials of the LEDs. The semiconductor structure is also one of the important factors that determines the light extraction by scattering effects produced by the surfaces (Zhmakin 2011).

Either the scattering effects can be produced with patterned/non-flattened surfaces or particles infused flattened outer surface. In a few cases, both of them are used to enhance the optical output. Few of them are patterned with particle coatings. However, the infinite pathways lead to the ultimate goal of achieving higher external output efficiency. In a few cases, the improved luminous efficiency and packing efficiency are achieved by scatterer-free methods (by using the bulky cube phosphors in micro-size) (Hu et al. 2013a). In white LEDs, the bulky micro-sized phosphor cubes are owing luminous and packing efficiencies of about 17% and 34% higher than that of the phosphor scatterer particles' infused conventional LEDs (Park et al. 2012). This is due to the reduction in scattering loss, because of the bulky micro-size phosphor with less grain boundaries (Park et al. 2013). Aside, using a scatterer infused layer to encapsulating the LED device will also affect the optical performance. This encapsulation packing technique, especially, using large refractive index encapsulant (Mont et al. 2008), is improving the light extraction efficiency more than the conventional packing techniques (Zheng et al. 2014). SiO_2 nanoparticles infused into the quantum dot/silicone encapsulation layer to enhance the scattering effect. The resultant white LED exhibits a higher luminous efficacy of 11.08% (Li et al. 2020).

3.2.1 SCATTERER-INFUSING/INDUCED ROUGHNESS METHODS

The optically active layers of the LEDs are doped with a few uniformly distributed scatterers, such as micro/nanostructured particles (spheres, rods, wires, etc.,) to enhance the external efficiency of the LEDs. The infused particles affect the refractive index. The change in refractive index causes changes in the wavelength of the optical

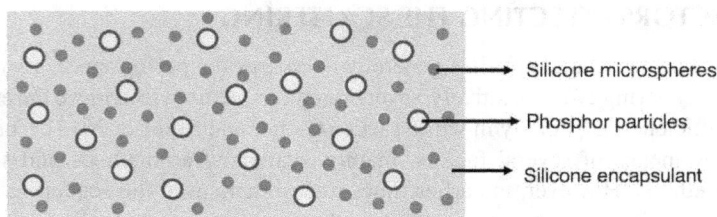

FIGURE 3.1 Schematic of the white LED encapsulant with the inclusion of silicone microspheres.

spectrum by the scattering effect. It is a sequential process of photon scattering and adsorption followed by conversion (Hu et al. 2013b; Wang et al. 2015). Specifically, photoluminescence phosphor is the major and common scatterer element infused in LEDs (Azarifar et al. 2021). The size (Sommer et al. 2010a) and concentration (Sommer et al. 2010b) of the spherical-shaped phosphor particles also plays a vital role. These factors further determine the optical performance of the resultant LEDs (Hu et al. 2012a). In few cases, the phosphor contents can be minimized by using micro-spheres, such as silicone micro-spheres in white LEDs as shown in Figure 3.1. The optical spectrum is experiencing a scattering process, due to the change in refractive index caused by the inclusion of micro-spheres. Yet, the minimization of phosphor particles (9.5% to 6.7%) with the inclusion of silicone microspheres (2.3%) in the white LED yields a similar colour temperature and emission flux decrement of 4.6% with a wide view angle (Kang et al. 2013). Roughness created on the surface of the active layer is highly influencing the scattering effect (Kawaguchi et al. 2007). In thin-film-based LED, micro-sized roughness surfaces are created on the active (n-GaN) layer of the LED device. The scattering occurring on this layer causes a significant effect of the optical performance of the LED (Li et al. 2017).

Besides, surface-functionalized metal-oxide particles are utilized as scatterers in LEDs, for instance, TiO_2 particles functionalized with silane. The more dispersibility is attained while the length of the silane carbon chain up to a certain length (from C_3 to C_6). This homogeneous dispersibility of the surface-functionalized TiO_2 particles offers colour uniformity and enhanced optical performance via the strong optical scattering (Song et al. 2021). In GaN-based LED, Ag nanoparticles are infused with SiO_2 on the p-GaN layer (current blocking layer). The infusion of Ag nanoparticles enhancing the light output power than the plain SiO_2 layer (Park et al. 2015). Apart from synthetic infusive materials as scatterers, bio-related materials are also used. For example, cocoon extracted silk fibroin protein-based hydrogel is placed as the lens material. This hydrogel consisting of the protein have crosslinked networks that causes scattering effect. In white LEDs, these silk hydrogel lenses provide a light extraction efficiency of 0.95 (Melikov et al. 2017).

3.2.1.1 Backscattering

Phosphor is a common scatterer particle infused into the LEDs (Lee et al. 2021). Even though the amount of phosphor particle infusion is less, the higher thickness is

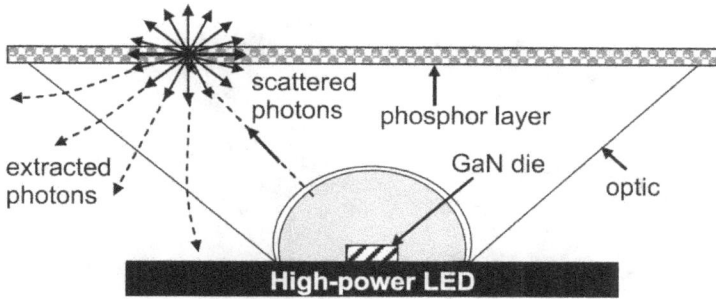

FIGURE 3.2 Illustration of the white LED package. (Reproduced with the copyright permission from (Narendran et al. 2005).)

helping to enhance the optical luminescence efficacy (23%) by reducing the trapping efficiency (Tran and Shi 2008). However, the backscattering of the optical spectrum by the absorption process tends to lessen the overall efficacy of white LEDs. This is due to the absorption of the backscattered photons by the LED components and further generate heat (Kim and Shin 2015). To extract the backscattered optical spectrum the phosphor layer is moved away from the source as shown in Figure 3.2. This increases the luminous efficacy up to 80 lm/W (Narendran et al. 2005). This enhances the improved light extraction, though, moving too far away will cause reduction in the light extraction (Liu et al. 2009b).

Apart from infusing additional scatterers, the phosphor particle itself is used with a different structure. Instead of spherical particles, rod structured nitride phosphor particles with less density and size are infused into the photonic crystal LEDs. The incident light is fragmented into several uniform parts due to the scattering effect of the rod-like nitride phosphor particles at the larger angles. The nitride phosphor possesses high refractive index, thus backward scattering power increases. Compared with the large sized particles, less sized particles can be able to scatter strongly (Li et al. 2015).

3.2.1.2 Bidirectional Scattering

Bidirectional scattering is the combination of both the forward scattering and the backward scattering (backscattering discussed earlier) of the optical spectrum as shown in Figure 3.3. Researchers developed several numerical models to predict the optical performance of white LEDs due to the bidirectional scattering occurred with inclusion of phosphor particle size, density, dispersion, thickness, scattering, absorption and optical conversion process (Hu and Luo 2012). Rather than infused as particles, phosphor is infused as a layer in the LED device to enhance the optical performance. The heat generation of the phosphor layer during the scattering process affects the optical performance of the white LEDs negatively (Luo and Hu 2014).

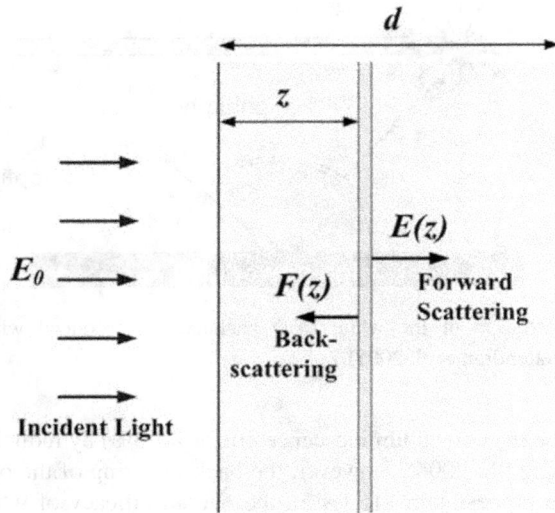

FIGURE 3.3 Schematic of the white LED package. (Reproduced with the copyright permission from (Luo and Hu 2014).)

3.2.1.3 Refractive Index Fluctuation Effect (Rayleigh Scattering) and Micro-Lens Effect

The Rayleigh scattering is generally produced with the fluctuations of refractive index in the presence of less-sized nanomaterials than that of the incident light wavelength (Byeon et al. 2015). In particular, the increment in refractive index increases the Rayleigh scattering. Metal-oxide nanoparticles (ZnO NPs) coated on a GaN-based LED surface, which is pre-coated with polystyrene (PS) hemisphere microstructures. This micro/nanohybrid structure consists of submicron particles (PS) roofed with nanoporous (ZnO) coating, correspondingly resulting inmicro-lensing and high refractive indexing (resulting in enhanced Rayleigh scattering) features. This collective effect increases the light extraction efficiency with enhanced optical output power of 77% (Mao et al. 2015).

3.2.1.4 Quantum Dot Scattering

The infusion of quantum dots (QD) in the LED will affect the optical performance. For instance, cadmium-selenium/zinc sulphide (CdSe/ZnS) quantum dots infused into a white LED and compared with the phosphor infused LED. Scattering and absorption in quantum dots are always different from the conventional particles used in LEDs. The infused less-stable QDs exhibit smaller radiant efficacy and weak scattering ability, due to the reabsorption behaviour. This finally tends the LED towards low optical performance (Li et al. 2018). Similarly, in the case of GaN-based ultra-violet (UV) LED, CdTe/C (cadmium-tellurium/carbon) hybrid quantum dots are encapsulated with NaCl (sodium chloride) crystals. These hierarchical particles are filled in the LED package along with the polymer layer. The metal side walls and the NaCl crystal grains enhance the strong scattering, which enhances the conversion efficiency. When considering the hierarchical particles possess various size distribution, the optical intensity caused via scattering also differed. This system offers an optical conversion efficiency of 72.6% with log-term stability and life-time (Hsu et al. 2015).

3.2.1.5 Fibre and Wire Structure Scattering

The optical performance of the LEDs can be improved by using single or multiple layered fibre, wire, and rod-like structures (Lee et al. 2015b). In order to improve the luminous efficiency of the QDs infused chip-on-board (COB) LEDs, a film layer of electrospunpolystyrenemicrofibres is introduced with different thicknesses. The microfibre infused QD COB LED possess improved and uniform optical scattering at different directions. At an optimized film thickness of 25 μm, the optical extraction is 31.7% in QDs infused COB LED (Liang et al. 2020). Similar to hybrid composite particles, the fibre layer can also be composited. In LED and OLED, a nanofibre layer of polyacrylonitrile is infused with SU-8 polymer to enhance the optical scattering, as shown in Figure 3.4. The change in the diameter of the nanofibre affects the optical performance of the resulting LEDs. Significantly, reduction in glare effect, parallel optical transmission and improvement in light extraction efficiency, are attained by tuning the diameter of the nanofiber (Lee et al. 2015a).

Similar to the nanofibres, nanowires also infused as a layer with the LED substrate. In a flexible white OLED, the Ag-nanowire layer is infused on the substrate of polyethylene terephthalate. The optical scattering effect induced by the silver nanowires improving the device efficiency (25%) and subsequently reducing the colour rendering index (Kim et al. 2018). The SiO_2 layer and the graphite nanoparticle layer are combinedly added to the mini full colour LED. The SiO_2 layer offers scattering effect that enhances the colour uniformity (65.9%). The additional graphite nanoparticle layer offers enhanced ambient contrast ratio (32.9%) through extinction effect. Also, the overall device performance improved by 168.8% (Li et al. 2022b). A meshed hybrid nanostructure (*p*-GaN/*p*-AlGaN) is also taken into account to use in deep UVLED. The structured hybrid mesh (*p*-GaN nanorod/*p*-AlGaN truncated nanocone) improves the light extraction efficiency for transverse magnetic polarized light (Zheng et al. 2019). Hydrothermally grown ZnO nanorods on the ITO substrate of GaN-based LED device is found to be enhancing the light extraction efficiency through scattering and wave-guide effects (Lee et al. 2011).

3.2.1.6 Inner Scattering Induced Waveguide Effect

Quantum dot colour convertors are mixed with mesoporous silica (SBA-15) as hybrid nanocomposite particles in white LEDs. The infusing of QDs into the pores of SBA-15 exhibits an inner scattering induced waveguide effect according to the incident light. The entire hybrid nanocomposite system (QD/SBA-15) possesses high refractive index. This combines with the inner scattering induced waveguide effect and enhances the light extraction efficiency (Li et al. 2021a).

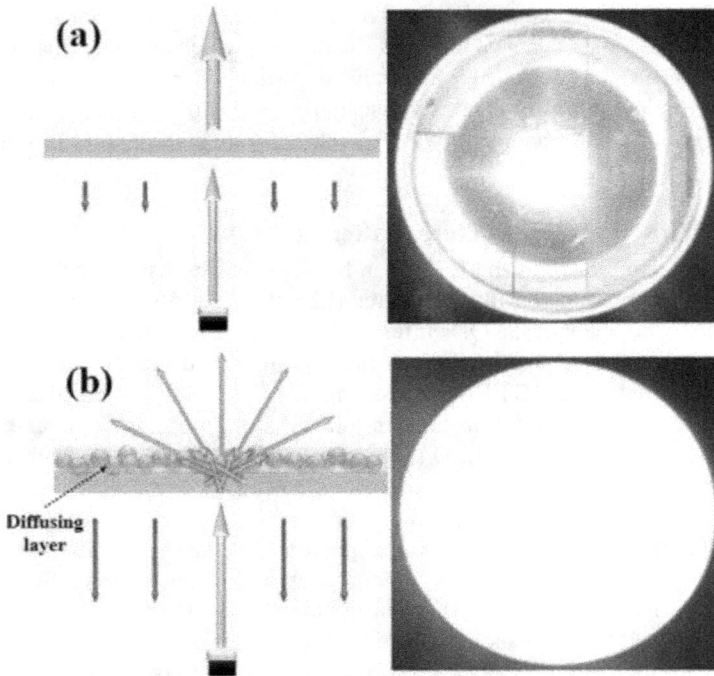

FIGURE 3.4 Illustration of the antiglare mechanism of the LED with (a) bare glass substrate and (b) nanofibre film. (Reproduced with the copyright permission from (Lee et al. 2015a).)

3.2.1.7 Aggregation-Induced Scattering

The addition of particles in quantum size (such as, QDs) in LEDs, even though their size effects are negligible, possibly influences the optical performance. The quantum dots tend to aggregate easily, because of their higher surface energy. When the concentration of the quantum dots increases, the aggregation also increased. For example, the CdSe/ZnS quantum dots are infused with the silicone matrix in white LED. The concentration increments of CdSe/ZnS quantum dots in silicone matrix improved the electromagnetic field distribution and backscattering strength as shown in Figure 3.5. This leads to reducing the external output of the LEDs. Compared with the positive influence of other effects, this aggregation-induced scattering effect is negatively influencing the optical performance of the LED devises (Li et al. 2021b).

3.2.1.8 Quantum Well-Surface Grating/Interface Effect (Surface Plasmon Generation)

Quantum wells are created with infusing the metal nanoparticles between the successive layers of LEDs to improve the external efficiency. The silver (Ag) nanoparticle is infused between wafer bond metal and GaN layer in the GaN-based vertical LED. This Ag-grating brings a quantum well, which induces the localized surface plasmon resonance. This enhances the output optical intensity by the surface

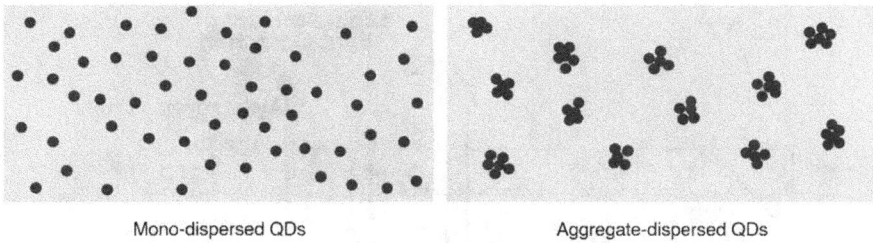

Mono-dispersed QDs   Aggregate-dispersed QDs

FIGURE 3.5 (a) Illustration of CdSe/ZnS quantum dots dispersed in silicone matrix (own image of the author).

plasmon-quantum well combined process (Lin et al. 2014). The Ag nanoparticle-infused OLED exhibits a light extraction efficacy of 82.5% is already reported (Al-Masoodi et al. 2020). Similarly, aluminium (Al) nanoparticles are infused with GaN-based deep-UV LED to create quantum well using formulation and numerical study. The interface scattering into the quantum walls and the surface plasmon effects are increasing the optical extraction efficiency (Chang et al. 2018). A gold (Au) nanoparticle influenced the InGaN/GaN blue LED, the inter-particles possess quantum wells. The quantum wells produce the localized surface plasmon resonance (absorb the photons and emits) along with the rough surface scattering process. Scattering minimizes the total internal reflection between the air and substrate layer. As shown in Figure 3.6, this combined effect enhances the light extraction in LEDs (Sung et al. 2009).

Similar to metal nanoparticles, metal/metal oxide core-shell nanoparticles are also employed in LEDs to produce the localized surface plasmon mode. This localized surface plasmon coupled with the polarized optical spectrum, hence enhances the optical extraction efficiency (Yang et al. 2020). The alloy nanoparticles prepared using gold and silver (Au-Ag) and infused as a layer in a perovskite LED. For an optimized particle size, alloy composition, and the functional layer thickness, the luminous efficiency is enhanced by 25%. This enhancement is due to the localized surface plasmon resonance wavelength (Zhang et al. 2019a). In GaN-based LED, the infusion of metal nanoparticles (Au and Ag) creates the quantum wells between the adjacent metal nanoparticles. This develops the localized plasmon resonance coupled with scattering effect. Jointly, they improve the performance of the LED device by 126% (Sung et al. 2010). Asides creating quantum wells between infused particles, they are created between the regularly grown gallium nitride nanocolumns. In blue LED, a light output efficiency of approximately 100% has been observed due to the scattering effect within the nanocolumns and the quantum wells (Tang et al. 2010).

3.2.1.9 Mie Scattering

Most of the LEDs infused with scatterer particles undergo Mie scattering. Cellulose nanocrystals infused white LEDs along with phosphor particles enhance the light output efficiency. This is due to the Mie theory of more light scattered by the scatterer particles (cellulose nanocrystals) and reached the phosphor particles. This

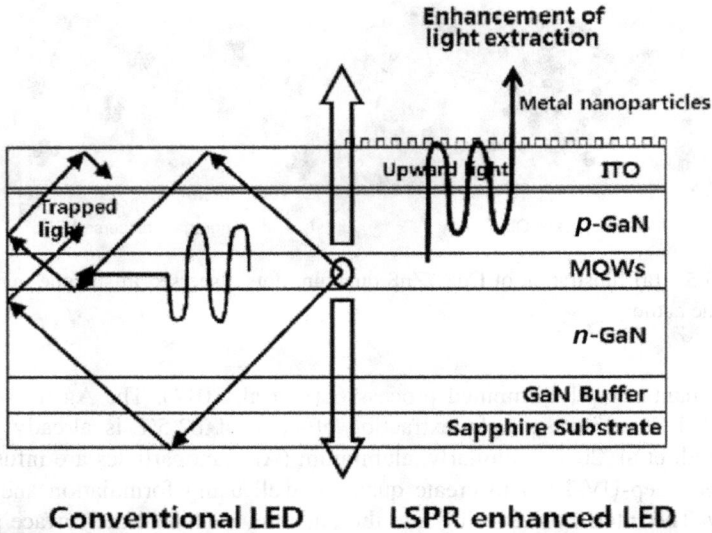

FIGURE 3.6 Illustration of localized surface plasmon resonance effect created by quantum wells in LED. (Reproduced with the copyright permission from (Sung et al. 2009).)

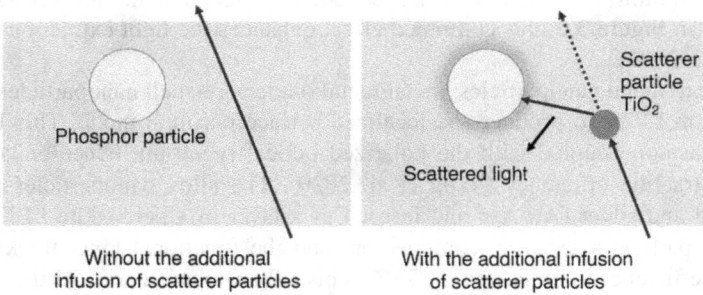

FIGURE 3.7 Scattering in wavelength conversion layer of conventional LED (left) and improved LED using Mie theory (right) (own image of the author).

further enhances the optical output efficiency (Chowdhury et al. 2021). Similarly, the polymer slabs of white LED are diffused with the TiO_2 particles. When the concentration of the TiO_2 particles increased, the photon travelling distance (i.e., mean-free path) became petite. This indicates that there is the possibility of blue light scattering in white LED than that of red light (Vos et al. 2013).

The wavelength alteration by changing the travel distance is the key of Mie scattering (Sommer et al. 2011). As shown in Figure 3.7, Mie theory is used in the white LEDs to reduce the quantity of phosphor requirement (up to 28%) with adding the scatterer (TiO_2) particles (Hsiao et al. 2013). Similarly, SiO_2 particles are infused with YAG:Ce phosphor in the white LED, which increases the scattering co-efficient

(Hu et al. 2012b). Likewise, multiple Mie scattering is achieved by the infusion of stacked layers of silica nanospheres in the blue LED. This enhances the optical output power to 2.7 times followed by the enhancement in optical extraction efficiency (Park et al. 2011). Two-size mixed phosphor particles are infused as the scatterer in the white LED. This hybrid scatterer system is increasing the angular colour uniformity, but slight dropping in the optical efficacy (Chen et al. 2014). However, the Mie scattering theory is only applicable for spherical-type scatterer particles.

3.2.1.10 Multiple Scattering

As discussed earlier, quantum well creation will induce the multiple scattering of light particles in LEDs in order to increase the external efficiency. Likewise, multiple scattering can be achieved by means of infusing phosphor. The phosphor is acting as a scatterer than that of an absorber. Hence, an increment in the phosphor concentration results in an increment in the scattering in the white LEDs (Leung et al. 2014). Nanoporous GaN is used in micro-LED to achieve multiple scattering with increased travel distance of photons. The strength of optical scattering enhanced by enhancing porosity and lessening the wavelength. This results in enhanced optical output (Kang et al. 2020).

3.2.1.11 Volumetric Scattering

The infusion of micro/nanoparticles with various shape, size, and fraction within the coating material coated on the surface of LED substrates is able to increase the optical extraction (Shiang and Duggal 2004). The phosphor particles infused with less size and high mass ratio results in influencing the scattering co-efficient and optical efficiency (Zhuo et al. 2019). The phosphor particles embedded in a blue OLED (phosphorescence OLED) providing excitation emission along with scattering. This results in enhancement in the external output efficiency (39 cd/A) in the phosphorescence OLEDs (Krummacher et al. 2006).

3.2.2 SCATTERER LAYER-ETCHED PATTERNING METHODS

In LEDs the high external quantum efficiency is accomplished with altering and/ or adding the optically active layer (generally, polymer film on the top of the LED configuration) in the LED. Introducing patterns of micro/nanostructures on the optically active layer via etching techniques are highly playing with the optical output of the LEDs. One or more effects can collectively provide the enhancement in the optical output in LED systems. In this method, the induced roughness affects the scattering effect strength by a few factors, such as angle of the roughness side-walls, scattering interface, size, inter-structure distance (periodicity), inclination, structures, cross-links, etc. (David 2013). The scattering caused by the patterned surface helps in improving the extracting the output light (Schad et al. 2004). While discussing about patterning/texturing on the surface of the LED, there are many structures and patterns that can be etched with different inter-structures and thicknesses. Each variation holds their own effects on the light extraction efficiency of the LEDs. Even the optically active layer coated with nanoparticle embedded systems can be etched for

Nanopatterned sapphire substrate

FIGURE 3.8   Schematic illustration the cross-sectional view of the nanopatterned sapphire substrate (own image of the author).

further optical output enhancement. The patterned sapphire substrate is providing the surface plasmon effect and total internal reflection compared with the pattern-free sapphire substrate, which enhances the light extraction efficiency of the LEDs (Zhang et al. 2019b).

A GaN-based LED is coated with resin, which is infused with zinc-oxide nanoparticles. This coating offers high refractive index on the LED substrate. Patterning on this layer is done with three different structures, such as non-etched, submicron-sized hole, submicron-sized pillars (with high-aspect-ratio) and microconvex arrays. The waveguided spectrum with optical scattering in submicron-sized pillars (with high-aspect-ratio) provides increased light extraction effect with a light output power of 19.6% (Byeon et al. 2015). The V-groove-like structures are etched on the upper surface of a multi-chip LED. The scattering effect occurred due to the etched V-groove pattern enhances the luminous efficiency by 9.6% with improved output power (Ding et al. 2017). In AlGaN-based deep ultraviolet (UV) LED, nanopattern-etched sapphire substrate is used as the scatterer layer, as shown in Figure 3.8. The nanopatterned substrate scatters more when compared with the pattern-free substrate. Meshed p-GaN layer along with the nanopatterned sapphire substrate is improving the light extraction efficiency by 112–124% (Zhang et al. 2021).

Similarly, scattering occurred in the interface between nanoporous patterned aluminium nitride layer and GaN layer reduces the non-radiative exciton recombination in GaN-based LED (Chen et al. 2009). An etched random nanorod structures of SU-8 polymer in OLED as the scattering layer, providing a rough surface for the scattering process. This gives an improvement in the external quantum efficiency (31%) with an enhancement in the viewing angle characteristics (Kwack et al. 2018). Likewise, sapphire substrate is patterned with nanocomb structures in GaN-based LED. This acts like a photonic crystal-like layer in order to produce the scattering effect. The scattering effect nanocomb layer is extracting more light particles. This increases the external quantum efficiency (43.7%) and optical output power (53.8%) (Liou et al. 2014).

3.2.2.1   Grating Effect (Bragg Scattering)

Periodically etched/patterned surfaces undergo the Bragg scattering effects in LEDs (Will et al. 2018). The pattern etching to make roughness on the photonic crystal surface offers enhancement in photon extraction via Bragg scattering in blue photonic crystal LEDs (PCLEDs). If the Bragg scattering is coherent, the scaping of photons from the rough surface happens, which results in improved external quantum efficiency of 30–50% compared with the conventional LEDs (Kim et al. 2005). In the photonic crystal-based red LED, the external quantum efficiency is increased from 30 to 120% due to the Bragg grating effect by introducing hole patterns on the photonic crystals. The LED emission of 642 nm is achieved via 0.5 distance ratio between the hole width and lattice remoteness (Kim et al. 2006). Periodically etched microstructures on the polymer films in organic LEDs (OLEDs) acts as Bragg grating. The Bragg scattering due to the surface etching enhances the polarization in the emission spectrum, which doubles the external efficiency of the LED without forfeiting the electrical properties (Lupton et al. 2000). Hence, periodicity is the key to produce Bragg scattering. This will increase the external efficiency with waveguide effect. However, Bragg scattering is able to develop the output light emission, depending on the angle of emission. The spectrum of light emission can be tuned with altering the distance between the micro-structures (i.e., Bragg grating) (Matterson et al. 2001).

3.2.2.2 Micro-/Nanolens Effect

In OLEDs, micro-lens arrays are printed with soft-lithography technique on a glass substrate. This micro-lens etching with a hemispherical structure diminishes the waveguiding in the glass substratum. This enhances the optical extraction of 85% without any changing in the electrical performance of the OLED. However, the finite size of the micro-lens arrays can obscure the image quality. Hence, this can be used for illumination applications rather than high-resolution displays (Peng et al. 2005). Patterning the sapphire substrate of the UVLED with micro-dome structures also causes the scattering effect. This improves the output of transverse magnetic polarized spectrum (Ooi and Zhang 2018). Similar to micro-lenses, nanosphere lenses are also fabricated on the substrate of the PCLED using lithographical etching technique. The nanospherical lenses created a diffused scattering effect along with the diffraction effect, which results in improved light extraction efficiency according to the improvement in the fill factor (Zhang et al. 2014a).

3.2.2.3 Quantum Well Effect

The etching technique is used to create quantum wells in GaN-based LED by patterning similar to the quantum wells created with infusing metal nanoparticles. Since the patterned quantum wells are in several numbers at a particular direction $1\bar{1}00$, are called multiple quantum wells. The light particles (photons) are able to scatter more into the air between the triangle-shaped walls than the vertical walls in the quantum wells. As shown in Figure 3.9, the air voids itself can act in scattering the optical spectrum (Okada et al. 2012). However, in both the cases, the optical extraction is improved with amended external quantum efficiency of 2% (Liu et al. 2009a).

3.2.2.4 Fresnel Scattering

As like the quantum well effect, the air gap with inclined metal side walls producing the Fresnel scattering. Due to the reflective scattering, the inclined metal side walls are able to enhance the light extraction efficiency. Zhang et al. tried to utilize this airgap with metal side walls and metal bottom walls to compare the light extraction efficiency. Relatively, the optical spectrum undergoes with total internal

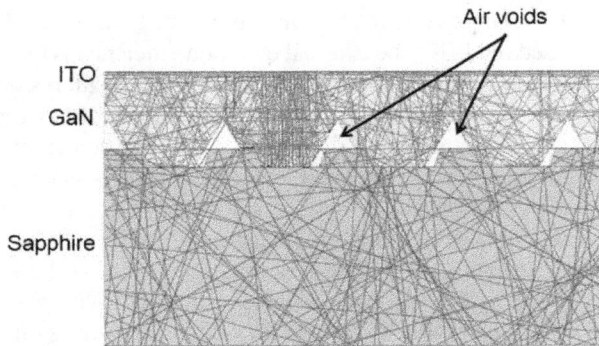

FIGURE 3.9　Air-voids (quantum wells) in GaN-based LED. (Reproduced with the copyright permission from (Okada et al. 2012).)

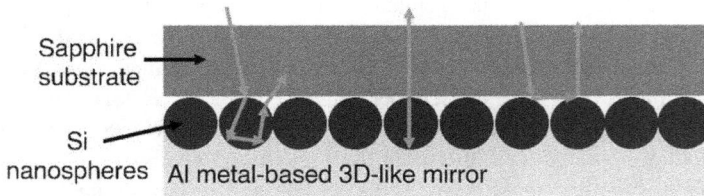

FIGURE 3.10　Schematic illustration of the metal side wall and metal bottom wall structures in deep UVLED (own image of the author).

reflection and Fresnel scattering in the model of airgaps with metal bottom walls. These combined effects enhance the light extraction efficiency of 48.3% (for deep UVLED with *p*-GaN contact layer) and 83.4% (for deep UVLED with *p*-AlGaN contact layer) (Zhang et al. 2017). Again, the same research group enhances the wall effect by using vertical metal side walls instead of inclined metal side walls. In this case, the vertical metal side walls in enhancing the light extraction efficiency in deep UVLED (Zhang et al. 2018). The Fresnel scattering is best utilized in GaN-based LEDs with patterning a back-side mirror (like 3D), as shown in Figure 3.10, offers an external quantum efficiency of 47.7% compared with the conventional one (Chen et al. 2021).

3.2.2.5　Coherent External Scattering

Coherent scattering occurs when the spontaneous scattering is absorbed using any particles or patterns in LED and re-scatter the optical spectrum stronger than the incident spectrum (Matsuta 2020). In LEDs, the thin film layer etched as two-dimensional photonic crystals on the edge-sides. The etched photonic crystals offer coherent external scattering around the non-textured surface as shown in Figure 3.11. This photonic crystal texturing enhances the photoluminescence (six-fold more than the non-textured one) by extracting (via absorption) the spontaneous emission occurred

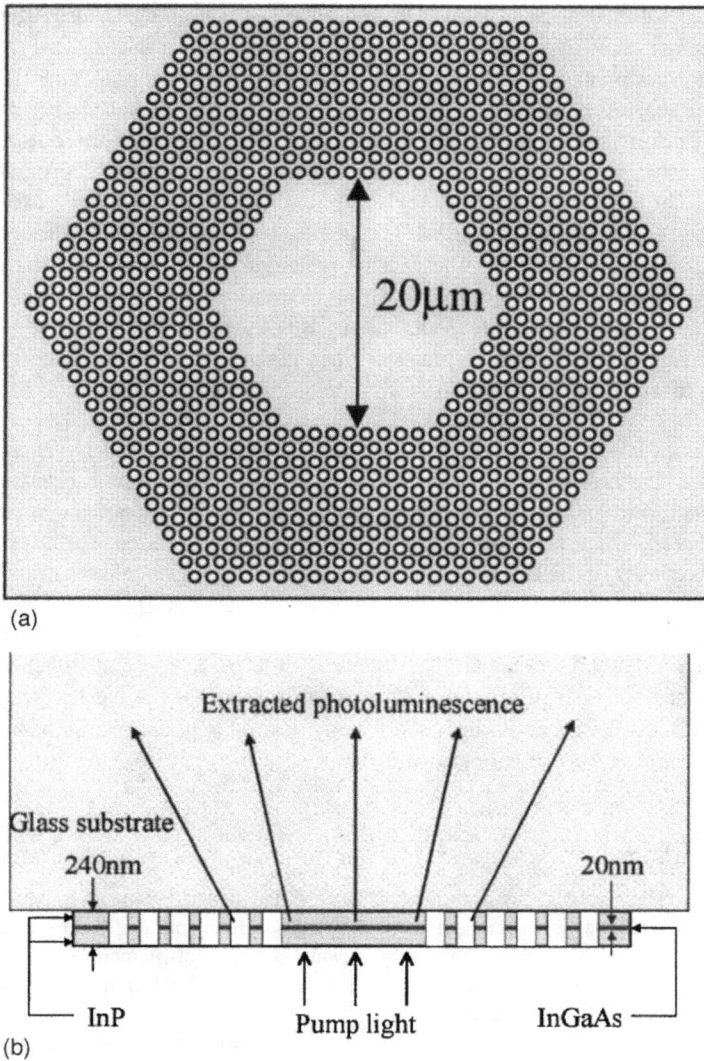

FIGURE 3.11 (a) Top view and (b) side view of the InGaAs/InP-based semiconductor thin film of the LED. (Reproduced with the copyright permission from (Boroditsky et al. 1999).)

in the non-textured area with an external quantum efficiency of 70% (Boroditsky et al. 1999).

3.3 CONCLUSION AND OUTLOOK

LEDs are the most-improved light source, which can replace the conventional light sources in various applications. This is achievable only by improving the optical performance of the various LEDs. The scattering effects in LEDs are affecting the optical performance negatively. But there are some techniques utilized to usefully convert the scattering effect to improve the optical performance. The discussions on both the scatterer-infusing/induced roughness methods and the scatterer layer-etched patterning methods are tending to increase or decrease the optical output (external) efficiency of the LEDs without losing their electrical properties/performance. The scatterer particles, such as phosphor, SiO_2, TiO_2, etc., are the most commonly used scatterer particles. The scatterers changing the refractive index of the optically active layer of the LED. This enhances the optical performances, such as external optical efficiency, light extraction efficiency, optical output power, etc., metal nanoparticles, such as Ag and Au are utilized to create quantum wells to offer localized surface plasmon resonance, which could couple with the scattering effect. This significantly enhances the optical performance of the LED device.

The loss of photons into the LED packages can be improved by encapsulating them with a scatterer infused layer. The internal loss of the light particles is due to the backscattering and the absorbing nature of the LED components can retrieve by using scatterer particles' infusion. The scatterers can also be used to make roughness on the active layer of the LED, which alters the path of the optical spectrum. Besides, patterned structures are carved via etching technologies that give waveguide effects. With respect to the periodicity and morphology of the roughness surface, various types of scattering effects occur. This increases the mean-free path of the photons. The rough surface also offers the uniformity in the output optical spectrum. Using scattering effects, LEDs could be tuned for any desired applications in the near future without sacrificing the optical performance.

REFERENCES

Al-Masoodi AHH, Goh BT, Nazarudin NFFB, Sarjidan MAM, Wong WS, Majid WHBA (2020) Efficiency enhancement in blue phosphorescent organic light emitting diode with silver nanoparticles prepared by plasma-assisted hot-filament evaporation as an external light-extraction layer. Mater Chem Phys 256:123618. https://doi.org/10.1016/j.matchemphys.2020.123618

Azarifar M, Cengiz C, Arik M (2021) Particle based investigation of self-heating effect of phosphor particles in phosphor converted light emitting diodes. J Lumin 231:117782. https://doi.org/10.1016/j.jlumin.2020.117782

Boroditsky M, Krauss TF, Coccioli R, Vrijen R, Bhat R, Yablonovitch E (1999) Light extraction from optically pumped light-emitting diode by thin-slab photonic crystals. Appl Phys Lett 75:1036–1038. https://doi.org/10.1063/1.124588

Byeon K, Cho J, Jo H, Lee H (2015) Fabrication of high-brightness GaN-based light-emitting diodes via thermal nanoimprinting of ZnO-nanoparticle-dispersed resin. Appl Surf Sci 346:354–360. https://doi.org/10.1016/j.apsusc.2015.03.022

Chang W-Y, Kuo Y, Yao Y-F, Yang CC, Wu Y-R, Kiang Y-W (2018) Different surface plasmon coupling behaviors of a surface Al nanoparticle between TE and TM polarizations in a deep-UV light-emitting diode. Opt Express 26:8340–8355. https://doi.org/10.1364/OE.26.008340

Chen D-C, Liu Z-G, Deng Z-H, Wang C, Cao Y-G, Liu Q-L (2014) Optimization of light efficacy and angular color uniformity by hybrid phosphor particle size for white light-emitting diode. Rare Met 33:348–352. https://doi.org/10.1007/s12598-013-0216-9

Chen L, Wang C, Huang J, Hong L-S (2009) A nanoporous AlN layer patterned by anodic aluminum oxide and its application as a buffer layer in a GaN-based light-emitting diode. Nanotechnology 20:085303. https://doi.org/10.1088/0957-4484/20/8/085303

Chen W, Niu J, Liu I, Wang Z, Cheng S, Liu W (2021) Study of a GaN-based light-emitting diode with a specific hybrid structure. ECS J Solid State Sci Technol 10:045001. https://doi.org/10.1149/2162-8777/abf0e7

Cho C, Zhao B, Tainter GD, Lee J, Friend RH, Di D, Deschler F, Greenham NC (2020) The role of photon recycling in perovskite light-emitting diodes. Nat Commun 11:611. https://doi.org/10.1038/s41467-020-14401-1

Chowdhury FI, Xu Q, Wang X (2021) Improving the light quality of white light-emitting diodes using cellulose nanocrystal-filled phosphors. Adv Photonics Res 2:2100006. https://doi.org/10.1002/adpr.202100006

Damulira E, Yusoff MNS, Sulaiman S, Zulkalfi NFH, Zulkalfi NA, Shakir NSA, Zainun MA, Omar AF, Taib NHM, Ali NKY (2020) Comparison of current–voltage response to diagnostic x-rays of five light-emitting diode strips. Appl Sci 10:200. https://doi.org/10.3390/app10010200

David A (2013) Surface-roughened light-emitting diodes: an accurate model. J Disp Technol 9:301–316. https://doi.org/10.1109/JDT.2013.2240373

Ding X, Tang Y, Li Z, Li J, Xie Y, Lin L (2017) Multichip LED modules with V-groove surfaces for light extraction efficiency enhancements considering roughness scattering. IEEE Trans Electron Devices 64:182–188. https://doi.org/10.1109/TED.2016.2628788

Fantini S, Franceschini MA, Fishkin JB, Barbieri B, Gratton E (1994) Quantitative determination of the absorption spectra of chromophores in strongly scattering media: a light-emitting-diode based technique. Appl Opt 33:5204–5213. https://doi.org/10.1364/AO.33.005204

Gather MC, Reineke S (2015) Recent advances in light outcoupling from white organic light-emitting diodes. J Photonics Energy 5:057607. https://doi.org/10.1117/1.JPE.5.057607

Gomard G, Preinfalk JB, Egel A, Lemmer U (2016) Photon management in solution- processed organic light-emitting diodes: a review of light outcoupling micro- and nanostructures. J Photonics Energy 6:030901.https://doi.org/10.1117/1.JPE.6.030901

Hart SJ, JiJi RD (2002) Light emitting diode excitation emission matrix fluorescence spectroscopy. Analyst 127:1693–1699. https://doi.org/10.1039/b207660h

Hsiao S-L, Hu N-C, Wu C-C (2013) Reducing the required amount of phosphor in warm white-light-emitting diodes by enhancing the scattering effect of wavelength conversion layer: a simulation study. Appl Phys Express 6:032102. https://doi.org/10.7567/APEX.6.032102

Hsu S, Chen Y, Tu Z-Y, Han H, Lin S, Chen T, Kuo H, Lin C-C (2015) Highly stable and efficient hybrid quantum quantum dot light-emitting diodes. IEEE Photonics J 7:1601210. https://doi.org/10.1109/JPHOT.2015.2487138

Hu R, Fu X, Zou Y, Luo X (2013a) A complementary study to "Toward scatter-free phosphors in white phosphor-converted light-emitting diodes:" comment. Opt Express 21:5071–5073. https://doi.org/10.1364/OE.21.005071

Hu R, Luo X (2012) A model for calculating the bidirectional scattering properties of phosphor layer in white light-emitting diodes. J Light Technol 30:3376–3380. https://doi.org/10.1109/JLT.2012.2218631

Hu R, Luo X, Feng H, Liu S (2012a) Effect of phosphor settling on the optical performance of phosphor-converted white light-emitting diode. J Lumin 132:1252–1256. https://doi.org/10.1016/j.jlumin.2011.12.059

Hu R, Luo X, Zheng H, Liu S (2012b) Optical constants study of YAG:Ce phosphor layer blended with SiO_2 particles by Mie theory for white light-emitting diode package. Front Optoelectron 5:138–146. https://doi.org/10.1007/s12200-012-0255-0

Hu R, Zheng H, Hu J, Luo X (2013b) Comprehensive study on the transmitted and reflected light through the phosphor layer in light-emitting diode packages. J Disp Technol 9:447–452. https://doi.org/10.1109/JDT.2012.2225021

Jin H, Chen L, Wu J, Wu Y, Zhu L, Li KH (2021) High-performance III-nitride light-emitting diode stripes. IEEE Tranctions Electron Devices 68:5629–5633. https://doi.org/10.1109/TED.2021.3108758

Kang J, Li B, Zhao T, Johar MA, Lin C, Fang Y-H, Kuo W-H, Liang K-L, Hu S, Ryu S-W, Han J (2020) RGB arrays for micro-LED applications using nanoporous GaN embedded with quantum dots. ACS Appl Mater Interfaces 12:30890–30895. https://doi.org/10.1021/acsami.0c00839

Kang YR, Kim KH, Kim WH, Jeon S, Jang MS, Kwak JS, Kim JP (2013) Utilization of silicone microspheres: improving color uniformity and reducing the amount of phosphor used in white light-emitting diodes. IEEE Trans Components, Pack Manuf Technol 3:1453–1457. https://doi.org/10.1109/TCPMT.2013.2261576

Kawaguchi Y, Nishizona K, Lee J-S, Katsuda H (2007) Light extraction simulation of surface-textured light-emitting diodes by finite-difference time-domain method and ray-tracing method. Jpn J Appl Phys 47:31–34. https://doi.org/10.1143/JJAP.46.31

Kim BS, Kim MK, Jo DS, Chae H, Cho SM (2018) Optimal structure of color-conversion layer for white organic light-emitting diode on silver-nanowire anode. ECS J. Solid State Sci. Technol. 7:R3176–R3181

Kim JH, Shin MW (2015) Thermal behavior of remote phosphor in light-emitting diode packages.IEEE Electron Device Lett 36:832–834. https://doi.org/10.1109/LED.2015.2441139

Kim T, Danner AJ, Choquette KD (2005) Enhancement in external quantum efficiency of blue light-emitting diode by photonic crystal surface grating. Electron Lett 41:1138–1139. https://doi.org/10.1049/el:20052643

Kim T, Leisher PO, Danner AJ, Wirth R, Streubel K, Choquette KD (2006) Photonic crystal structure effect on the enhancement in the external quantum efficiency of a red LED. IEEE Photonics Technol Lett 18:1876–1878. https://doi.org/10.1109/LPT.2006.881235

Kioupakis E, Rinke P, Delaney KT, Walle CG Van De (2011) Indirect Auger recombination as a cause of efficiency droop in nitride light- emitting diodes. Appl Phys Lett 98:161107. https://doi.org/10.1063/1.3570656

Krummacher BC, Choong V, Mathai MK, Choulis SA, So F, Jermann F, Fiedler T, Zachau M (2006) Highly efficient white organic light-emitting diode. Appl Phys Lett 88:113506. https://doi.org/10.1063/1.2186080

Kumar R, Adhikari S, Chatterjee V, Pal S (2021) Recent advances and challenges in AlGaN-based ultra-violet light emitting diode technologies. Mater Res Bull 140:111258.https://doi.org/10.1016/j.materresbull.2021.111258

Kwack JH, Choi J, Park CH, Hwang H, Park YW, Li B-K (2018) Simple method for fabricating scattering layer using random nanoscale rods for improving optical properties of organic light-emitting diodes. Sci Rep 8:14311. https://doi.org/10.1038/s41598-018-32538-4

Lee HJ, An S, Hwang JH, Jung S, Jo HS, Kim KN, Shim YS, Park CH, Yoon SS, Park YW, Ju B (2015a) Novel composite layer based on electrospun polymer nanofibers for efficient light scattering. ACS Appl Mater Interfaces 7:68–74. https://doi.org/10.1021/am5075387

Lee KH, Moon Y, Song J, Kwak JS (2015b) Light interaction in sapphire/MgF2/Al triple-layer omnidirectional reflectors in AlGaN-based near ultraviolet light-emitting diodes. Sci Rep 5:9717. https://doi.org/10.1038/srep09717

Lee S, Kim D, Kang K, Lee E, Jo Y, Jung S, Oh JT, Jeong H, Seong T (2021) Using self-aligned Si barrier to enhance the contrast ratio and performance of pixelated light emitting diode for vehicle headlamp. ECS J Solid State Sci Technol 10:045003. https://doi.org/10.1149/2162-8777/abf47d

Lee Y-S, Jung Y-I, Noh B-Y, Park I-K (2011) Emission pattern control of GaN-based light-emitting diodes with ZnO nanostructures. Appl Phys Express 4:112101. https://doi.org/10.1143/APEX.4.112101

Leung VYF, Lagendijk A, Tukker TW, Mosk AP, IJzerman WL, Vos WL (2014) Interplay between multiple scattering, emission, and absorption of light in the phosphor of a white light-emitting diode. Opt Express 22:8190–8204. https://doi.org/10.1364/OE.22.008190

Li J, Chen J, Lin L, Li Z, Tang Y, Yu B, Ding X (2015) A detailed study on phosphor-converted light-emitting diodes with multi-phosphor configuration using the finite-difference time-domain and ray-tracing methods. IEEE J Quantum Electron 51:1–10. https://doi.org/10.1109/JQE.2015.2472341

Li J, Tang Y, Li Z, Ding X, Li Z (2017) Study on the optical performance of thin-film light-emitting diodes using fractal micro-roughness surface model. Appl Surf Sci 410:60–69. https://doi.org/10.1016/j.apsusc.2017.03.041

Li J, Tang Y, Li Z, Ding X, Rao L, Yu B (2018) Effect of quantum dot scattering and absorption on the optical performance of white light-emitting diodes. IEEE Trans Electron Devices 65:2877–2884. https://doi.org/10.1109/TED.2018.2830798

Li J, Tang Y, Li Z, Li J, Ding X, Yu B, Yu S, Ou J, Kuo H (2021a) Toward 200 lumens per watt of quantum-dot white-light-emitting diodes by reducing reabsorption loss. ACS Nano 15:550–562. https://doi.org/10.1021/acsnano.0c05735

Li X, Chen J, Liu Z, Deng Z, Huang Q, Huang J, Guo W (2022a) (Ce, Gd):YAG-Al$_2$O$_3$ composite ceramics for high-brightness yellow light-emitting diode applications. J Eur Ceram Soc 42:1121–1131. https://doi.org/10.1016/j.jeurceramsoc.2021.11.027

Li Z, Li J, Deng Z, Liang J, Li J (2021b) Unraveling the origin of low optical efficiency for quantum dot white light-emitting diodes from the perspective of aggregation-induced scattering effect. IEEE Trans Electron Devices 68:1738–1745. https://doi.org/10.1109/TED.2021.3060698

Li Z, Li J, Li J, Deng Z, Deng Y, Tang Y (2020) Scattering effect on optical performance of quantum dot white light-emitting diodes incorporating SiO$_2$ nanoparticles. IEEE J Quantum Electron 56:3600109.https://doi.org/10.1109/JQE.2020.2986018

Li Z, Wu J, Ren Z, Song Y, Li J (2022b) Improving ambient contrast ratio and color uniformity of mini full color light-emitting diodes using an SiO$_2$/graphite bilayered packaging structure. J Electron Packag 144:011010. https://doi.org/10.1115/1.4050202

Liang G, Yu S, Tang Y, Lu Z, Yuan Y, Li Z, Li J (2020) Enhancing luminous efficiency of quantum dot-based chip-on-board light-emitting diodes using polystyrene fiber mats. IEEE Trans Electron Devices 67:4530–4533. https://doi.org/10.1109/TED.2020.3014061

Lin C, Hsieh C, Tu C, Kuo Y, Chen H, Shih P, Liao C, Kiang Y, Yang CC, Lai C, He G, Yeh J, Hsu T (2014) Efficiency improvement of a vertical light-emitting diode through surface plasmon coupling and grating scattering. Opt Express 22:842–856. https://doi.org/10.1364/OE.22.00A842

Liou J, Chen C, Chou P, Tsai Z, Chang Y, Liu W-C (2014) Implementation of a high-performance GaN-based light-emitting diode grown on a nanocomb-shaped patterned

sapphire substrate. IEEE J Quantum Electron 50:973–980. https://doi.org/10.1109/
    JQE.2014.2365022

Liu Y, Yen C, Yu K, Chen T, Chen L, Tsai T-H, Liu W-C (2009a) Characteristics of a low-damage
    GaN-based light-emitting diode using a KOH-treated wet-etching approach characteristics
    of a low-damage GaN-based light-emitting diode using a KOH-treated wet-etching
    approach. Jpn J Appl Phys 48:082104.https://doi.org/10.1143/JJAP.48.082104

Liu Z, Liu S, Wang K, Luo X (2010) Measurement and numerical studies of optical properties
    of YAG:Ce phosphor for white light-emitting diode packaging. Appl Opt 49:247–257.
    https://doi.org/10.1364/AO.49.000247

Liu Z, Liu S, Wang K, Luo X (2009b) Optical analysis of phosphor's location for high-power
    light-emitting diodes. 9:65–73. https://doi.org/10.1109/TDMR.2008.2010250

Luo X, Hu R (2014) Calculation of the phosphor heat generation in phosphor-converted light-
    emitting diodes. Int J Heat Mass Transf 75:213–217. https://doi.org/10.1016/j.ijheatm
    asstransfer.2014.03.067

Lupton JM, Matterson BJ, Samuel IDW, Jory MJ, Barnes WL, Lupton JM, Matterson BJ, Samuel
    IDW, Jory MJ, Barnes WL (2000) Bragg scattering from periodically microstructured
    light emitting diodes Bragg scattering from periodically microstructured light emitting
    diodes. Appl Phys Lett 77:3340–3342. https://doi.org/10.1063/1.1320023

Mao P, Mahapatra A, Chen J, Chen M, Wang G, Han M (2015) Fabrication of polystyrene/
    ZnO micronano hierarchical structure applied for light extraction of light-emitting
    devices. ACS Appl Mater Interfaces 7:19179–17188. https://doi.org/10.1021/acs
    ami.5b04911

Matsuta H (2020) Coherent forward-scattering spectra of Cs I 852.1-nm with a light-emitting
    diode and a diode laser in a Voigt configuration. Spectrochim Acta Part B 171:105935.
    https://doi.org/10.1016/j.sab.2020.105935

Matterson BBJ, Lupton JM, Safonov AF, Salt MG, Barnes WL, Samuel IDW (2001) Increased
    efficiency and controlled light output from a microstructured light-emitting diode. Adv
    Mater 13:123–127. https://doi.org/10.1002/1521-4095(200101)13:2<123::AID-ADMA
    123>3.0.CO;2-D

Melikov R, Press DA, Kumar BG, Dogru IB, Sadeghi S, Chirea M, Yılgör İ, Nizamoglu S
    (2017) Silk-hydrogel lenses for light-emitting diodes. Sci Rep 7:7258. https://doi.org/
    10.1038/s41598-017-07817-1

Mont FW, Kim JK, Schubert MF, Schubert EF, Siegel RW (2008) High-refractive-index TiO_2-
    nanoparticle-loaded encapsulants for light-emitting diodes. J Appl Phys 103:083120.
    https://doi.org/10.1063/1.2903484

Mukherjee A, Deyasi A, Das B (2016) Effect of electron-electron scattering on spontaneous
    emission for bulk semiconductor and light emitting diode. Adv Ind EngManag 5:165–
    171. https://doi.org/10.7508/aiem.2016.01.031

Narendran N, Gu Y, Freyssinier-Nova JP, Zhu Y (2005) Extracting phosphor-scattered photons
    to improve white LED efficiency. Phys Stat Sol 202:60–62. https://doi.org/10.1002/
    pssa.200510015

Okada N, Uchida K, Miyoshi S, Tadatomo K (2012) Green light-emitting diodes fabricated on
    semipolar (11–22) GaN on r-plane patterned sapphire substrate. Phys Status Solidi A
    209:469–472. https://doi.org/10.1002/pssa.201100385

Ooi YK, Zhang J (2018) Light extraction efficiency analysis of flip-chip ultraviolet light-
    emitting diodes with patterned sapphire substrate. IEEE Photonics J 10:8200913. https://
    doi.org/10.1109/JPHOT.2018.2847226

Park HK, Oh JH, Do YR (2012) Toward scatter-free phosphors in white phosphor-converted
    light-emitting diodes. Opt Express 20:25593–25601. https://doi.org/10.1364/
    OE.20.010218

Park HK, Oh JH, Do YR (2013) Toward scatter-free phosphors in white phosphor-converted light-emitting diodes: reply to comments. Opt Express 21:5074–5076. https://doi.org/10.1364/OE.21.005074

Park J, Han J, Seong T (2015) Improving the output power of GaN-based light-emitting diode using Ag particles embedded within a SiO$_2$ current blocking layer. Superlattices Microstruct 83:361–366. https://doi.org/10.1016/j.spmi.2015.03.027

Park YJ, Kang JH, Kim HY, Lysak VV, Chandramohan S, Ryu JH, Kim HK, Han N, Jeong H, Jeong MS, Hong C-H (2011) Enhanced light emission in blue light-emitting diodes by multiple Mie scattering from embedded silica nanosphere stacking layers. Opt Express 19:23429–23435. https://doi.org/10.1364/OE.19.023429

Peng H, Ho YL, Yu X, Wong M, Kwok H (2005) Coupling efficiency enhancement in organic light-emitting devices using microlens array – theory and experiment. J Disp Technol 1:278–282. https://doi.org/10.1109/JDT.2005.858944

Pyo B, Joo CW, Kim HS, Kwon B-H, Lee J-I, Lee J, Suh MC (2016) A nanoporous polymer film as a diffuser as well as a light extraction component for top emitting organic light emitting diode with strong microcavity structure.Nanoscale 8:8575–8582. https://doi.org/10.1039/C6NR00868B

Ryu H, Shim J (2010) Structural parameter dependence of light extraction efficiency in photonic crystal InGaN vertical light-emitting diode structures. IEEE J Quantum Electron 46:714–720. https://doi.org/10.1109/JQE.2009.2035933

Schad S, Neubert B, Eichler C, Scherer M, Habel F, Seyboth M, Scholz F, Hofstetter D, Unger P, Schmid W, Karnutsch C, Streubel K (2004) Absorption and light scattering inInGaN-on-sapphire- and AlGaInP-based light-emitting diodes. J Light 22:2323–2332. https://doi.org/10.1109/JLT.2004.832437

Shedbalkar A, Andreev Z, Witzigmann B (2016) Simulation of an indium gallium nitride quantum well light-emitting diode with the non-equilibrium Green's function method. Phys Status Solidi B 253:158–163. https://doi.org/10.1002/pssb.201552276

Shedbalkar A, Witzigmann B (2018) Non equilibrium Green's function quantum transport for green multi-quantum well nitride light emitting diodes. Opt Quantum Electron 50:1–10. https://doi.org/10.1007/s11082-018-1335-1

Shiang JJ, Duggal AR (2004) Application of radiative transport theory to light extraction from organic light emitting diodes. J Appl Phys 95:2880–2888. https://doi.org/10.1063/1.1644037

Sommer C, Reil F, Krenn JR, Hartmann P, Pachler P, Hoschopf H, Wenzl FP (2011) The impact of light scattering on the radiant flux of phosphor-converted high power white light-emitting diodes. J Light Technol 29:2285–2291. https://doi.org/10.1109/JLT.2011.2158987

Sommer C, Wenzl FP, Reil F, Krenn JR, Hartmann P, Pachler P, Tasch S (2010a) On the effect of light scattering in phosphor converted white light-emitting diodes. In: Advanced Photonics and Renewable Energy, OSA Technical Digest (CD). p SOTuB5

Sommer C, Wenzl FP, Reil F, Krenn JR, Pachler P, Tasch S, Hartmann P (2010b) A comprehensive study on the parameters effecting color conversion in phosphor converted white light emitting diodes. In: Tenth International Conference on Solid State Lighting. p 77840D

Song GY, Jang I, Jeon S, Ahn S, Kim J-Y, Kim S young, Sa G (2021) Controlling the surface properties of TiO$_2$ for improvement of the photo-performance and color uniformity of the light-emitting diode devices. J Ind Eng Chem 94:180–187. https://doi.org/10.1016/j.jiec.2020.10.031

Sung J, Kim B, Choi C, Lee M, Lee S, Park S, Lee E, Beom-Hoan O (2009) Enhanced luminescence of GaN-based light-emitting diode with a localized surface plasmon resonance. Microelectron Eng 86:1120–1123. https://doi.org/10.1016/j.mee.2009.01.009

Sung J, Yang JS, Kim B, Choi C, Lee M, Lee S, Park S, Lee E, OB-H (2010) Enhancement of electroluminescence in GaN-based light-emitting diodes by metallic nanoparticles. Appl Phys Lett 96:261105. https://doi.org/10.1063/1.3457349

Tang T, Lin C, Chen Y, Shiao W, Chang W, Liao C, Shen K, Yang C, Hsu M, Yeh J, Hsu T (2010) Nitride nanocolumns for the development of light-emitting diode. IEEE Trans Electron Devices 57:71–78. https://doi.org/10.1109/TED.2009.2034795

Taniyasu Y, Kasu M, Makimoto T (2006) An aluminium nitride light-emitting diode with a wavelength of 210 nanometres. Nature 441:325–328. https://doi.org/10.1038/natu re04760

Tran NT, Shi FG (2008) Studies of phosphor concentration and thickness for phosphor-based white light-emitting-diodes. J Light Technol 26:3556–3559. https://doi.org/10.1109/ JLT.2008.917087

Vos WL, Tukker TW, Mosk AP, Lagendijk A, IJzerman WL (2013) Broadband mean free path of diffuse light in polydisperse ensembles of scatterers for white light-emitting diode lighting. Appl Opt 52:2602–2609. https://doi.org/10.1364/AO.52.002602

Wang J, Lo JCC, Lee SWR, Yun F, Tao M (2015) Modeling and parametric study of light scattering, absorption and emission of phosphor in a white light-emitting diode. In: ASME 2015 International Technical Conference and Exhibition on Packaging and Integration of Electronic and Photonic Microsystems collocated with the ASME 2015 13th International Conference on Nanochannels, Microchannels, and Minichannels. pp IPACK2015-48664, V002T02A046

Will P, Schwarz EB, Fuchs C, Scholz R, Lenk S, Reineke S (2018) Scattering quantified: Evaluation of corrugation induced outcoupling concepts in organic light-emitting diodes. Org Electron 58:250–256. https://doi.org/10.1016/j.orgel.2018.04.016

Yang Y, Sun H, Zhang Y, Su H, Shi X, Guo Z (2020) Surface plasmon coupling with radiating dipole for enhancing the emission efficiency and light extraction of a deep ultraviolet light emitting diode. Plasmonics 15:881–887. https://doi.org/10.1007/s11468-019-01107-4

Yue Q, Li K, Kong F, Zhao J, Liu M (2016) Analysis on the effect of amorphous photonic crystals on light extraction efficiency enhancement for GaN-based thin-film-flip-chip light-emitting diodes. Opt Commun 367:72–79. https://doi.org/10.1016/j.opt com.2015.12.072

Zhang J, Change L, Zhao Z, Tian K, Chu C, Zheng Q, Zhang Y, Li Q, Zhang Z-H (2021) Different scattering effect of nano-patterned sapphire substrate for TM- and TE-polarized light emitted from AlGaN-based deep ultraviolet light-emitting diodes. Opt Mater Express 11:729–739. https://doi.org/10.1364/OME.416605

Zhang Y, Meng R, Zhang Z, Shi Q, Li L, Liu G, Bi W (2017) Effects of inclined sidewall structure with bottom metal air cavity on the light extraction efficiency for AlGaN-based deep ultraviolet light-emitting diodes. IEEE Photonics J 9:1600709. https://doi.org/ 10.1109/JPHOT.2017.2736642

Zhang Y, Sun H, Zhang S, Li S, Wang X, Zhang X, Liu T, Guo Z (2019a) Enhancing luminescence in all-inorganic perovskite surface plasmon light-emitting diode by incorporating Au-Ag alloy nanoparticle. Opt Mater (Amst) 89:563–567. https://doi.org/10.1016/j.opt mat.2019.01.074

Zhang Y, We iT, Xiong Z, Chen Y, Zhen A, Shan L, Zhao Y, Hu Q, Li J, Wang J (2014a) Enhancing optical power of GaN-based light-emitting diodes by nanopatterning on indium tin oxide with tunable fill factor using multiple-exposure nanosphere-lens lithography. J Appl Phys 116:194301.https://doi.org/10.1063/1.4901829

Zhang Y, Wei T, Xiong Z, Shang L, Tian Y, Zhao Y, Zhou P, Wang J, Li J (2014b) Enhanced optical power of GaN-based light-emitting diode with compound photonic crystals by multiple-exposure nanosphere-lens lithography. Appl Phys Lett 105:013108.https://doi. org/10.1063/1.4889745

Zhang Y, Zhang J, Zheng Y, Sun C, Tian K, Chu C, Zhang Z-H, Liu JG, Bi W (2019b) The effect of sapphire substrates on omni-directional reflector design for flip-chip near ultra-violet light-emitting diodes. IEEE Photonics J 11:8200209. https://doi.org/10.1109/ JPHOT.2018.2889319

Zhang Y, Zheng Y, Meng R, Sun C, Tian K, Geng C, Zhang Z, Liu G, Bi W (2018) Enhancing both TM- and TE-polarized light extraction efficiency of AlGaN-based deep ultraviolet light-emitting diode via air cavity extractor with vertical sidewall. IEEE Photonics J 10:8200809. https://doi.org/10.1109/JPHOT.2018.2849747

Zheng H, Li L, Lei X, Yu X, Liu S, Luo X (2014) Optical performance enhancement for chip-on-board packaging leds by adding TiO$_2$/silicone encapsulation layer. IEEE Electron Device Lett 35:1046–1048. https://doi.org/10.1109/LED.2014.2349951

Zheng Y, Zhang Y, Zhang J, Sun C, Chu C, Tian K, Zhang Z, Bi W (2019) Effects of meshed p-type contact structure on the light extraction effect for deep ultraviolet flip-chip light-emitting diodes. Nanoscale Res Lett 14:149. https://doi.org/10.1186/s11671-019-2984-0

Zhmakin AI (2011) Enhancement of light extraction from light emitting diodes. Phys Rep 498:189–241. https://doi.org/10.1016/j.physrep.2010.11.001

Zhuo N, Zhang N, Jiang T, Chen P, Wang H (2019) Effect of particle sizes and mass ratios of a phosphor on light color performance of a green phosphor thin fi lm and a laminated white light-emitting diode. RSC Adv 9:27424–27431. https://doi.org/10.1039/c9ra05503g

4 Challenges in Fabrication and Packaging of LEDs

Nesa Majidzadeh and Hossein Movla

CONTENTS

4.1 LIGHTNING: HISTORY AND CURRENT STATUS

Lightning is one of the most significant modern inventions. It has a variety of industrial applications in addition to residential and public lighting. The introduction of
real, energy-efficient lighting in the home and business has a big impact on contem-
porary life. Lighting technology has advanced quickly over the past few decades,
starting with the mass production of incandescent light bulbs and continuing with
the development of fluorescence lights. Artificial lighting has evolved since Edison's
invention of the incandescent bulb in 1879 [1] to enhance power output and reduce
system size. Innovation in lighting and display was made possible by the creation of
commercially available visible red light-emitting diodes (LEDs) in the 1960s at the
General Electric company, which utilized Holonyak's groundbreaking research, and
later at Monsanto and Hewlett Packard [2] and blue LEDs in 1993 [3]. Due to their
performance becoming more and more competitive with that of conventional light
sources, as seen in Figure 4.1, they have recently gained a lot of attention [4].

LEDs have become the emerging device in many applications across all industries, from simple lights to sophisticated electrical processes. Due to their exceptional
qualities, including environmental friendliness, compact size, temperature control,
extended lifespan, low power consumption, and low cost, LEDs have drawn a lot
of attention. LEDs have a service life of between 25,000 and 100,000 hours [4].

DOI: 10.1201/9781003340577-4

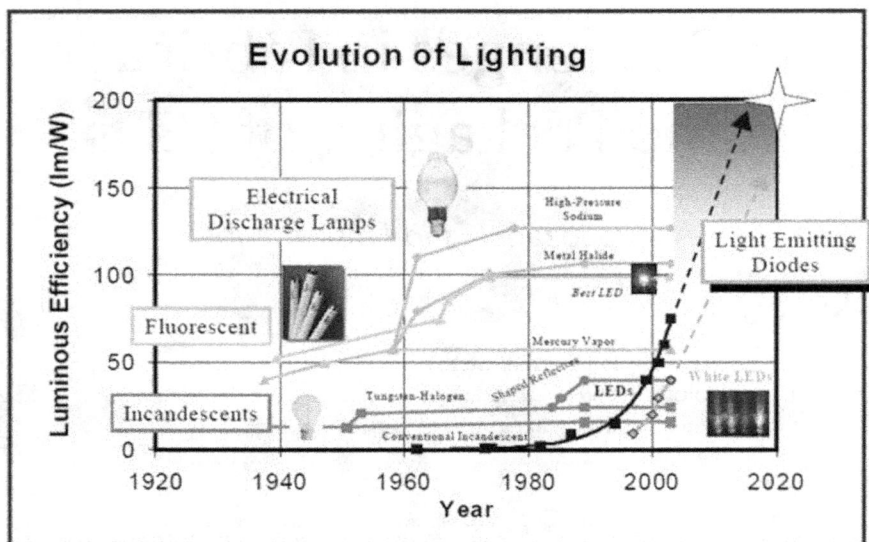

FIGURE 4.1 Schematic comparison of projected and actual LED performance with that of traditional light sources [4]. (Reprinted from [4] with permission from Elsevier.)

Additionally, LEDs have great conversion efficiency, converting about 80% into light energy and the other 20% into heat, which results in energy savings [5].

Compared to traditional lighting sources, LED lights have a number of benefits, including a very long lifespan, a high level of resilience, low energy consumption, and the ability to produce a variety of colors, which makes them stand out [6,7]. In addition to offering white light in a range of color temperatures and producing millions of different colors, LED lighting fixtures are fully dimmable and can be managed by either a basic switch or complex optical feedback driver electronics that balance the red, green, and blue (RGB) light output over temperature [7].

4.2   LED TECHNOLOGY TRENDS

Fossil fuels currently account for the majority of the world's energy usage. Consequences of the use of fossil fuels include climate change, the greenhouse effect, and other phenomena. Since the manufacturing of the luminaires and lamps that enable this uses just a small portion of the total energy consumed, the lighting industry may significantly contribute to sustainability. About 20% of all energy is required for lighting to illuminate various indoor and outdoor spaces. Lighting, for example, accounts for 18% of energy consumption by end uses in the residential sector in the EU and 30% in the USA [8]. According to estimates, lighting accounts for 18–21% of total energy consumption, including 2.5–3.5% for public and road lighting and 15%–18% for indoor lighting, which includes trade and services (7%), housing (8%), and industrial (3%). As a result, indoor lighting analysis is crucial to the sustainability of energy. In the United States, the use of electrical energy for lighting applications is predicted to decrease by 15%, 40%, and up to 75% in the years 2020, 2030, and 2035, respectively, when traditional lighting sources are replaced with LEDs [8, 9, 10].

Interior lighting solutions rank among the most significant ways to save energy in the majority of industrialized nations due to the rising cost of electricity. As a result, the history of indoor lighting is not solely dependent on the creation of new light sources that contribute to sustainability and energy efficiency. Energy conservation is becoming a growing global trend and LED adoption has skyrocketed. For sustainable development, finding an energy-saving lighting solution is therefore imperative, and LED technology is key.

The rapid advancement of LED technology, the world's energy supply, as well as environmental concerns, have all contributed to the rapid expansion of LED applications and the anticipated continued rapid growth in these areas [7]. For instance, only in the United States, do fluorescent lamps require 32 tons of mercury annually. Lighting consumes 30% of all electricity used in the United States, which is the largest category, and 22% of all oil, gas, coal, and nuclear energy used in the country [9]. The efficiency of white LEDs has now surpassed that of the most energy-efficient fluorescent lighting, allowing for the anticipated rapid penetration of LED lighting into both the $70 billion global general lighting market as well as the specialty lighting market.

Although LED lighting is anticipated to someday replace conventional lighting, the most common uses for colorful and white LEDs are [11, 12] as follows:

- car exterior and interior and lighting;
- backlighting for small and middle-sized liquid crystal displays (LCD);
- single and traffic lighting; and
- mini and micro-LED display
- high-power LED lasers and projectors
- general lightning
- medical equipment

Those uses today make up nearly 90% of the demand for LEDs. Traffic signals, billboards, exterior car lighting, and cell phone display backlights are just a few of the many-colored illumination applications that already use LEDs. The wide application of LEDs in our life is shown in Figure 4.2.

In mobile and relatively low-wattage applications, white lighting applications are also growing quickly, but high-wattage general illumination applications are more difficult to penetrate, mostly due to cost reasons. Micro-light-emitting diodes (μ-LEDs), which have great brightness, longevity, resolution, and efficiency qualities, have recently propelled advancements in display technology.

OLED panels undoubtedly have benefits like the best black level and great brightness, but there is also competition from emerging technologies. When it comes to price and performance, OLED is being challenged by both micro-LED and mini-LED. Micro-LEDs, like OLED, use self-emitting technology, which is the reason they are still an extremely expensive niche product [13–16].

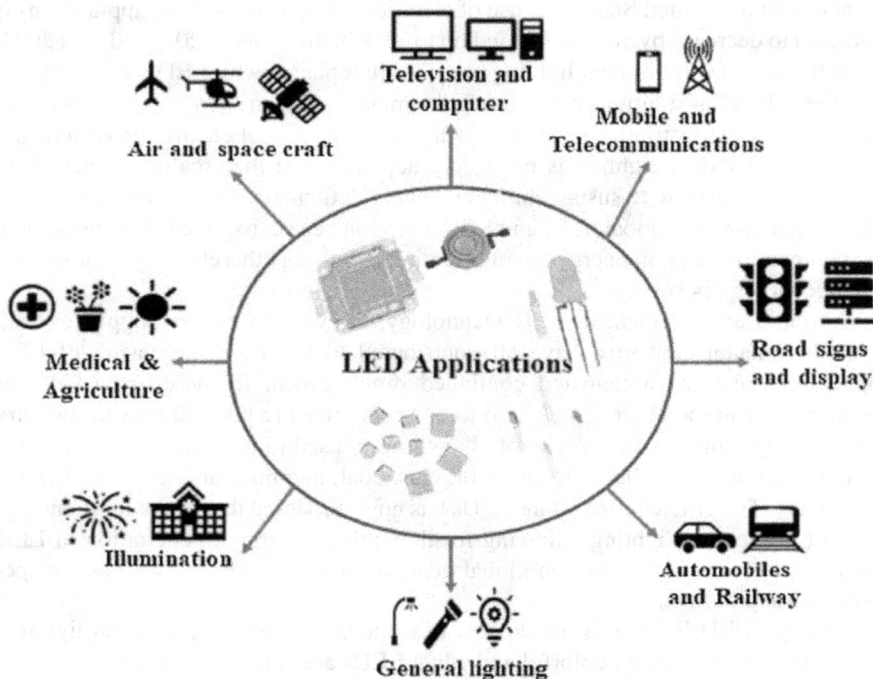

FIGURE 4.2 LED application in our life [4]. (Reprinted from [4] with permission from Elsevier.)

The demand for and technological advancements in LEDs have evolved. Scientists are developing methods to enhance it. To advance electronic technology, high power (HP) density and high-temperature capability are required.

4.3 TECHNOLOGICAL ADVANCEMENTS IN LED FABRICATION

The efficiency claims from various LED manufacturers have climbed to what seem to be impossibly high numbers as a result of tremendous advancements in die, phosphor, and package technology during the previous few years [7, 17–18]. Epitaxial growth, chip production, packaging, and lamp integration are examples of technical processes used to make LED lamps. The first three processes are crucial for optical efficiency because they determine the internal quantum efficiency (IQE), light extraction efficiency, and packaging efficiency, respectively. White LED efficiency is typically expressed in lumens per watt (lm/W). The majority of white, mid-power LEDs employ a die to produce blue light, which is afterward changed by a phosphor system into white light. Manufacturers work to improve the die and phosphor of an LED to increase its efficiency. The efficiency with which blue dies produce blue light from electrical energy has been increasing. Blue dies have climbed the efficiency curve year after year, even though it is practically difficult to be 100% efficient at doing this [17].

Larger and larger dies can be placed in mid-power packages and driven at relatively modest currents to boost the efficiency of LEDs. Because blue dies used in mid-power parts frequently achieve their best efficiencies at lower driving currents, this improves overall efficiency [8, 19–21]. Increasing the die area in mid-power LEDs is one trend that has emerged in recent years.

Even though this may appear like an expensive technique to increase efficiency, die costs have been falling very quickly as a result of an oversupply in the worldwide LED market. Therefore, utilizing larger dies truly doesn't present a major cost disadvantage. Making blue dies more efficient from the start would be a more elegant and economical method to improve them, allowing manufacturers to employ less expensive, smaller dies [17, 20].

The vertical layer-by-layer atom deposition dominates epitaxial growth, which is a typical nanomanufacturing technique [20–23]. Both the lateral overgrowth and the micromachining for substrate patterning are associated with temperature cycles and mechanical deformation during this process, which could result in dislocations and other flaws. Additionally, atomic, molecular, and macroscale understandings of this bottom-up process are required. Microstructures like microlenses, which may enhance light extraction, are used in chip production. Internal quantum efficiency and extraction efficiency are the two factors that combine to make up LED performance, which increased from 0.1 lumens per watt in 1970 to 100 lm/W in 1999 [24].

LEDs' capacity to replace conventional lighting systems grows as they become more efficient. Additionally, fewer LEDs are required to generate the same amount of light, which lowers the price of the bulbs and fixtures. The fact that a more efficient LED generates less waste heat means that the heatsinks in bulbs and fixtures can also be decreased, which is an extra benefit. All of this simplifies the task of the lighting designer and increases the commercial appeal of LED lighting.

4.4 LED PACKAGING

The key challenges in the packaging of LEDs are depicted in Figure 4.3. Similar to how computers enabled innovative packages during the previous 25 years, consumer electronics applications are also driving advancements in packaging technology [17]. Consumer electronics applications can be found on bigger devices like televisions and smart appliances as well as smaller ones like smartphones, game consoles, radios, and music players. Electronics that are made in very big volumes are becoming smaller, faster, and less expensive. Since the characteristics and performance of LEDs are greatly influenced by their packaging, LED packaging has received more attention over the past few decades. To protect the LED chip from direct contact with the environment, LED packages to encase the chip with the LED. Wafer mounting, dicing, die or chip attachment, wire bonding, and encapsulation are only a few of the several procedures involved in producing an LED package [17].The packaging process includes coatings, polymer layers, layers of phosphors, and structures that may have lenses or reflectors. Figure 4.4 depicts the process flow diagram for the typical LED packaging process [4, 8].

A typical phosphor-based high-power white LED package is made up of an LED chip, thermal and electrical connections (die attach and wire bond material),

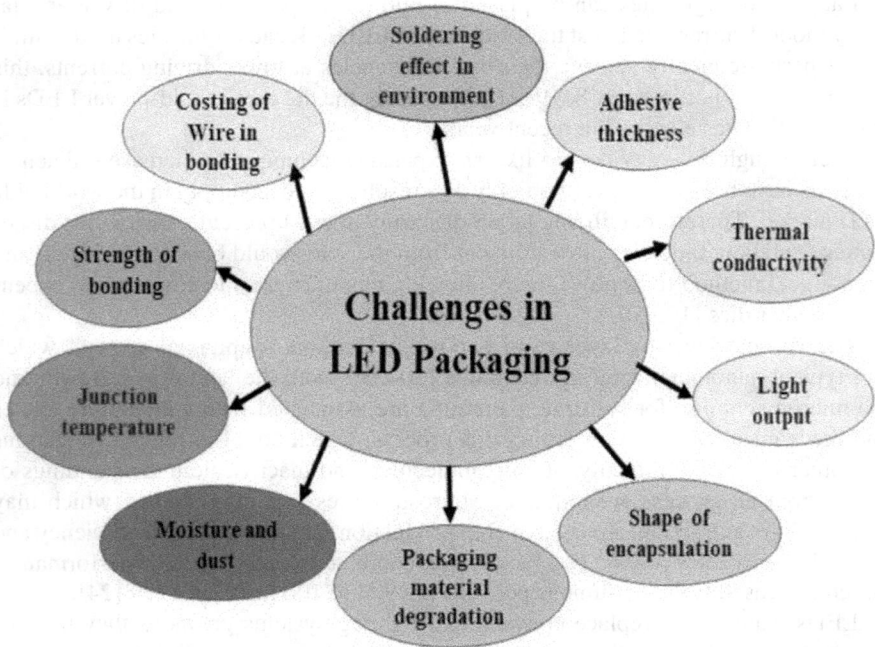

FIGURE 4.3   Challenges in LED packaging [4]. (Reprinted from [4] with permission from Elsevier.)

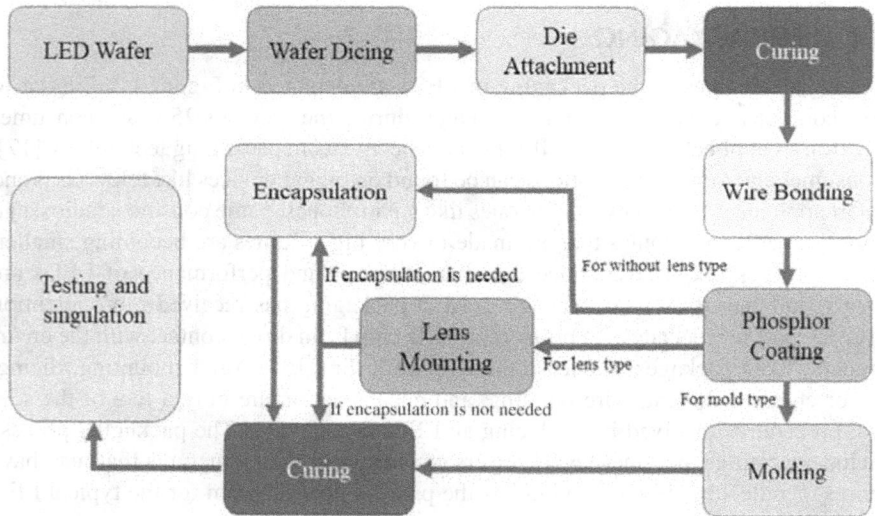

FIGURE 4.4   LED packaging process flow chart [4]. (Reprinted from [4] with permission from Elsevier.)

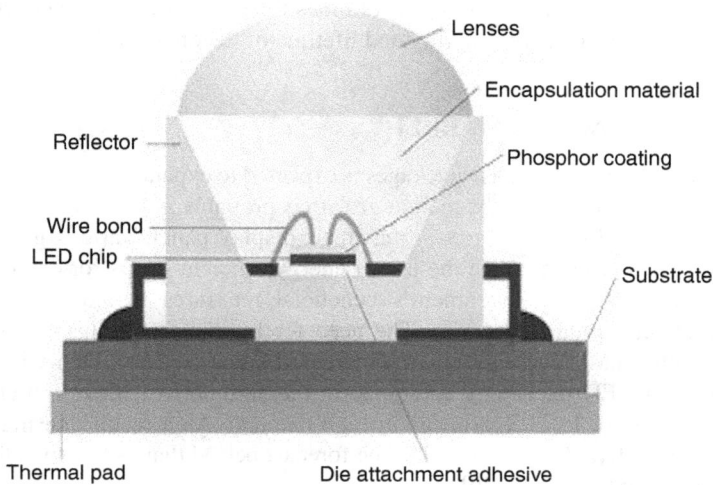

Lenses

Encapsulation material

Reflector

Phosphor coating

Wire bond

LED chip

Substrate

Thermal pad   Die attachment adhesive

FIGURE 4.5 Schematic cross-sectional view of white LED packages. (Reprinted from [4] with permission from Elsevier.)

a substrate, a phosphor-containing encapsulating material, an optional optical lens, and an optional reflector cup [25]. In this scenario, the performance of the LEDs is greatly influenced by the optical design and the encapsulating material for the LED packaging that contains phosphor. A typical high-power white LED package structure is depicted in Figure 4.5 and consists of a silicone encapsulant combined with a phosphor material, a leadframe with a reflector cup, an LED die, and a die attach adhesive, gold wires, and the LED itself [17, 21].

The following is a description of the packing process [17]:

(1) the lead frame is cleaned and baked before use;
(2) a conductive die attach adhesive is used to secure the blue LED chip to the leadframe's reflector cup in the middle;
(3) the adhesive is cured at least for 45 minutes at 175°C;
(4) the LED die is electrically connected to the leadframe using wire-bonding;
(5) the leadframe's reflector cup is filled with a silicone encapsulant that has been blended with a YAG:Ce^{3+} phosphor.
(6) the optical lens is attached to the top surface of the encapsulant to make the with-lens type package.

To increase the luminescence efficiency of LED chips, exterior leads are connected to the electrode by the LED packaging. The package defines the performance and is a component of the electronic system. To fit this modern society, LED packaging must be stable and dependable [7]. As previously said, throughout the complete LED value chain, LED packaging has the highest market value and profit margin. Similar to integrated circuit (IC) device packaging, the quality of packaging and assembly materials as well as their processing determines the yield of the device packing and assembly as well as the dependability and lifetime of the device [17].

4.5 LED PACKAGING MARKET

The market for high-power LED packages is expected to expand due to rising demand in lighting applications. Shortly, market growth is probably going to be fueled by the rising demand for LED packages in the smart display panel. Various governments have enacted laws to encourage the use of energy-efficient LEDs for environmental reasons. Therefore, the government's beneficial measures would probably have a positive effect on market growth. The need for Internet of Things (IoT)-enabled lighting fixtures and smart lighting has increased significantly, which will increase the demand for LED packaging solutions. In the automotive industry, high-quality interior and exterior LED lighting is in high demand. As a result, the market will experience considerable growth during the forecast period thanks to expanding automotive lighting applications [26].

The consumer electronics industry's rising need for smart LEDs is what's fueling the expansion of the LED packaging market. In addition, the need for effective UV LED packages for disinfection systems has increased due to the global COVID-19 pandemic. The increased reliability of UV-C LEDs' ability to kill germs and viruses has increased the demand for LED packaging under pandemic conditions [26].

The main drivers of the global market demand are the rising demand for effective UV LED packages, the increasing demand for LED packages in the market for display panels, as well as the rising number of government initiatives and regulations promoting the use of LEDs for energy efficiency and environmental benefits. For instance, Statista reports that the global market for LED lighting was valued at USD 83.86 billion in 2021 and is expected to increase to USD 160.03 billion by 2026 [26]. As a result, the global market growth is being boosted by the rising demand for LED packaging as a result of the growing use of LED lighting.

But over the projected period of 2022–2028, the market's expansion will be hampered by the absence of open, universal standards and by the market's high saturation from the presence of numerous manufacturers. Additionally, it is projected that during the forecast period, the rising preference for UV LED-based curing systems for print label and packaging solutions as well as the expanding acceptance of CSP LEDs in automotive lighting applications would serve as catalysts for the market demand [26].

4.6 MATERIALS AND PACKAGING CHALLENGES

Although the basic functionalities of the materials used to package LEDs, such as die bonding, wire bonding, and die encapsulation, are shared with those used to package IC devices, there are also additional needs specific to the packaging of LEDs [17]. The encapsulant materials are the best example of how this is done. With silica filling, encapsulant materials for IC devices can have low coefficients of thermal expansion (CTE), excellent levels of flame resistance, and minimal moisture absorption.

However, due to requirements for optical transmission, the encapsulant materials for LEDs cannot be filled with regular silica fillers [17].

All potential outcomes for the package after it leaves the factory are taken into account when determining the commercial electronic system's reliability. Optical, thermal, electrical, surfaces/interfaces diffusion, deformation/stress, and testing processes all take place simultaneously [23, 24]. Temperature, thermal cycling, and electrical current (electromigration) are examples of environmental stresses that cause package degradation. These pressures are often externally imposed and cannot be regulated in commercial applications due to cost considerations. The following details how problems with temperature cycles, mechanical shock/drop, and electromigration might influence the dependability of consumer goods.

4.7 LIGHT EXTRACTION

Numerous elements affect an LED package's reliability. The heat generated inside the packaging is one of the most crucial factors. Low light extraction, which is the key factor in LED degradation, decreased light output, and shorter lifetime, is the source of the majority of the heat that is produced (Figure 4.6). The heat problem can be resolved using a variety of methods. Increasing light extraction is a practical solution to this issue [27]. High light output and small package size are made possible by a single big chip mounting. However, as the die size grows, chip quantum efficiencies decrease considerably, mostly because the chip's sidewalls emit less light. The surface area for heat flow and light extraction is greatly increased when several little chips are used as opposed to a single large chip.

The result is a significantly improved light efficacy and superior thermal performance due to the many tiny chip mounting. The complexity, size, and cost of LED packaging could all rise as a result. It is still difficult to create multi-chip LED packages that are both small and inexpensive. Epoxy encapsulants with high refractive indices and excellent transparency at target wavelengths can help improve light extraction.

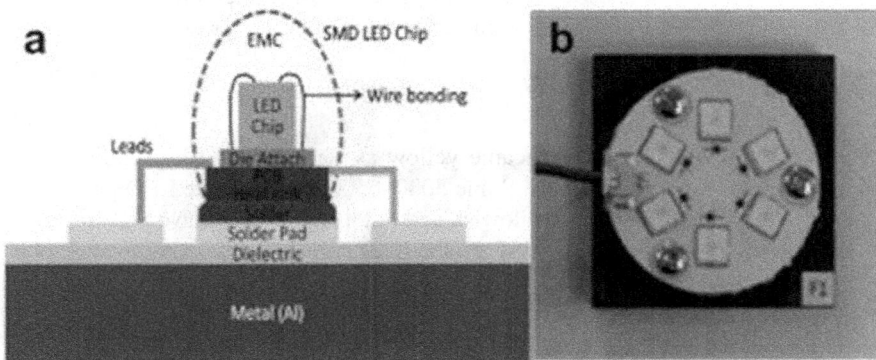

FIGURE 4.6 General LED package for heat conduction [17]. (Reprinted from [17] with permission from Elsevier.)

The extraction efficiency is significantly influenced by the encapsulant's absorption coefficient [28].

4.8 THERMAL YELLOWING

High-power LEDs have been constrained by reliability and efficiency issues. Up until recently, the packaging materials were not being challenged by the high-power LEDs' continuous advances in light output. In theory, driving larger chips with more current should increase the amount of light produced [29]. However, the majority of high-power LEDs only turn 15% of the energy they receive into light, with the remaining energy being lost as heat [7, 17]. High junction temperatures brought on by high-power LEDs make thermal management a crucial packaging concern. For the packaging of high-power LEDs, it is important to maximize heat removal for safe junction temperature functioning and minimize thermal stresses brought on by the CTE mismatch of materials. Because of this, standard epoxy systems are typically used to encase LEDs. These systems have a tendency to break down quickly when exposed to high temperatures or intense ultraviolet radiation, which results in the intrusion of harmful moisture and air as well as discoloration. High-power white and blue LEDs have short lives mostly because of these degradations [7].

There are three different causes for the yellowing of led panel lights [5, 29]:

1. Yellowing caused by the material change of the light guide plate itself.

This is the most prevalent and serious yellowing issue on the market, as well as the problem with our actual yellowing quality. The middle of the panel light's light-emitting surface turns yellow when you turn on this ultra-thin panel light. This issue is caused by the polystyrene (PS) material light guide plate.

In petrochemicals, it is a relatively minor clear brittle glue. It only has a specific gravity of 1.047–1.049. The light transmittance is 87% when it is completely transparent. It is a plastic with the second-lowest refractive index after polycarbonate, where PS is the next-most transparent substance. The PS light guide plate will cost significantly less than a PMMA light guide plate, but it is much more prone to weather damage, such as turning yellow, and extreme cold brittleness.

2. The LED chips are abnormal (mainly phosphor problems causing yellowing).

Some panel lights at that time became yellow as a result of LED technology and illicit businesses making shortcuts. Some 5000–6500 K cold white LED panel lights to begin to generate light that is yellowish and reddish after they have been turned on [30]. The primary causes of this issue are the LED chip packaging method and phosphor consumption ratio.

3. The external use environment allows the stains adhering to the light guide plate to infiltrate into the light guide plate and create yellowing:

The diffuser plate, which is the top layer, serves as both a homogeneous light guide and a barrier against dirt and dust from contaminating the light guide plate.

In practice, however, many users will disregard the impact of the external environment on the lamp's internal structure, particularly in some unique use situations like chemical factories, laboratories, and restrooms that frequently have water vapor present. The particular air elements at these locations have a propensity to slowly infiltrate the lamp body through the side of the panel light, contaminating the light guide plate and resulting in yellowing. Or, for instance, the surface of the panel light will turn yellow due to the diffuser's long-term degradation by water vapor.

4.9   UV RADIATION-INDUCED YELLOWING

Another challenging issue for the degradation and aging mechanism of LEDs is photodegradation of encapsulants induced by UV radiation from LED dies and outdoor. In LED packaging, several materials and polymers are used and their applications are related to the cost of the production, power consumption, and efficiency of the LED.

The most widely applied material is silicones with compounds like polydimethylsiloxanes, trimethylsilyloxy, poly diphenyl siloxanes, methyl phenyl, and terminated silicone. The latter has a special structure since it combines organic groups (vinyl, methyl, etc.) with an inorganic Si-O backbone to form a semi-organic material [31]. The silicones and siloxanes have some special features, such as better biocompatibility, great moisture/UV light resistance, and structural stability during service [32–37]. The fact that silicones retain their electrical capabilities at high temperatures and in humid settings makes them a desirable option for LED applications [34]. The acknowledged drawbacks of silicones include their generally low tear strength, poor adhesion, low glass transition temperature, high CTE, high moisture and gas penetration, and their often-lengthy production periods.

Bisphenol-A polycarbonate (BPAPC), a thermoplastic with the components depicted in Figure 4.7 is one of the most frequently utilized polymers in LED device packaging. BPA-PC is widely used for secondary optics in LED devices because of its high fracture resistance, relatively high elastic modulus, and excellent thermal stability.

Polymethyl methacrylate (PMMA) is another thermoplastic that is widely utilized in LED lenses and packaging [38–40]. Figure 4.7 depicts the material's structural elements. One of the least expensive alternatives on the market is PMMA. PMMA's widespread use is due in part to its glass transition temperature, T_g, which is between 105 and 110°C and facilitates processing and production. Albeit it is the cheapest material used in LED packaging, it has very low UV resistance [41, 42].

Due to a favorable combination of low cost, ease of manufacture, mechanical stabilities, and moisture resistance [43, 44], epoxy materials hold the largest proportion of the LED packaging industry [45, 46, 47]. Epoxies are frequently utilized in printed circuit boards as die-attach and underfill adhesives (PCBs). As LED encapsulants, epoxy resins are recognized to have two significant drawbacks. The first problem is the resin's brittleness as a result of excessive crosslinking. The photodegradation of epoxy under radiation at high temperatures is the other drawback.

Massive chain scission occurs under the aforementioned circumstances, causing a rapid discoloration. Since discoloration decreases transparency and light production, it cannot be allowed in LEDs.

4.10 STRESS/DELAMINATION

During temperature cycling in the manufacturing process, a CTE mismatch between the bonded parts and the bonding solder produces pressures that might result in delamination between the bonded surfaces. Inappropriate solder and process management can occasionally cause a short circuit in the device. Solders have a rather high degree of wettability, which might allow them to overflow a specific area of contacts and cause a short. It is common knowledge that shrinkage and internal stress develop as epoxy resin cure. In reality, internal stress increases with increasing resin and substrate material thermal expansion coefficient differences, which could lead to processing failures or decreased LED reliability.

4.11 RELIABILITY AND LIFETIME

The LED die must be attached to a heat sink or substrate, frequently via solder, to disperse the heat produced during operation. Hot spots that emerge from an inadequate thermal route caused by voids in the solder connect will eventually cause thermal runaway and failure [17, 48]. Typically occurring close to the bonded surface between the solder and the heat sink, whisker development brought on by electromigration—which can be triggered by internal strain, temperature, humidity, and material characteristics—can result in electrical short circuits.

The following factors should be taken into account when selecting a die to attach material [17]:

(1) stress relaxation at the interface;
(2) excellent adhesion between the bonded surfaces;
(3) efficient heat dissipation and high thermal conductivity; and
(4) CTE matching materials between the bonded surfaces.

Encapsulant, wire, and phosphor failure in a package are all possible. Overheated epoxy encapsulant causes wire-bond breakage or separation and die-attach strength reduction. The chip and epoxy eventually delaminate as a result of these issues. Lead wire mechanical stress, which can result in open circuits inside the device, is another failure mode. At normal operating temperature, improper lead wire soldering pressure, position, and direction can build up tension and cause the leads to bend toward the LED's body.

REFERENCES

[1] T.A. Edison, US223898 A (1880).
[2] S. Nakamura, MRS Bulletin, 34, 101 (2009).

[3] S. Nakamura, T. Mukai, and M. Senoh, Applied Physics Letters, 64, 1687 (1994).

[4] Md. Abdul Alim, M.Z. Abdullah, M.S. Abdul Aziz, R. Kamarudin, Die attachment, wire bonding, and encapsulation process in LED packaging: A review, Sensors and Actuators A: Physical, 329, 112817 (2021).

[5] T.H. Chiang, Y.C. Lin, Y.F. Chen, E.Y. Chen, Effect of anhydride curing agents, imidazoles, and silver particle sizes on the electrical resistivity and thermal conductivity in the silver adhesives of LED devices, J. Appl. Polym. Sci. 133(26),1–9(2016).

[6] J. Piprek, Nitride Semiconductor Devices, Principles and Simulation (Wiley, Berlin, 2006).

[7] S. Nakamura, G. Fasol, S.J. Pearton, The Blue Laser Diode: The Complete Story (Springer, 2000 Edition, 2nd updated and extended Ed.).

[8] Md. Abdul Alim, M.Z. Abdullah, M.S. Abdul Aziz, R. Kamarudin, Die attachment, wire bonding, and encapsulation process in LED packaging: A review, Sensors and Actuators A 329, 112817(2021).

[9] Navigant Consulting Inc, Energy Savings Forecast of Solid-State Lighting in General Illumination Applications for the U.S. Department of Energy, no. August, p. 68, 2014.

[10] F.G. Montoya, A. Peña-García, A. Juaidi, F. Manzano-Agugliaro, Indoor lighting techniques: an overview of evolution and new trends for energy saving, Energy Build. 140, 50–60(2017).

[11] E. Fred Schubert, Light-Emitting Diodes (Cambridge: Cambridge University Press, 2nd ed., 2006).

[12] BMW company.

[13] MSI company.

[14] Y. Huang, E.L. Hsiang, M.Y. Deng, et al. Mini-LED, micro-LED and OLED displays: present status and future perspectives. Light Sci Appl. 9, 105 (2020).

[15] R. Zhu, Z. Luo, H. Chen, Y. Dong, S.-T. Wu. Realizing Rec. 2020 color gamut with quantum dot displays. Opt Express. 23(18), 23680–23693 (2015).

[16] Lumileds Company. Webpage available at https://lumileds.com/company/blog/driving-high-efficiency-in-mid-power-leds/ (2022).

[17] J-.K. Sim, K. Ashok, Y-.H. Ra, H-.C Im, B-.J Baek, C.-R. Lee, Characteristic enhancement of white LED lamp using low temperature co-fired ceramic-chip on board package, Current Applied Physics, 12(2), 494–498 (2012).

[18] D. Lu, C.P. Wong (eds.), Materials for Advanced Packaging, Springer US. Yuan-Chang Lin, Yan Zhou, Nguyen T. Tran, Frank G. Shi (Authors) Chapter 18. LED and Optical Device Packaging and Materials, 629–680.

[19] I. Corporation, From Sand to Silicon 'Making of a Chip' Illustrations 32nmHigh-K / Metal Gate–Version, no. April, 2011, 1–16.

[20] K.S. Siow, Die-attach Materials for High Temperature Applications in Microelectronics Packaging: Materials, Processes, Equipment, and Reliability, 2019.

[21] J.P. You, Y. He, F.G. Shi, Thermal management of high power LEDs: impact ofdie attach materials, Proc. Tech. Pap.–2007 Int. Microsystems, Packag. Assem. Circuits Technol. Conf. IMPACT239–242 (2007).

[22] Sheng Liu, Zhiyin Gan, Xiaobing Luo, Kai Wang, Xiaohui Song, Zhaohui Chen, Han Yan, Zongyuan Liu, Pei Wang, Wei, Multi-physics multi-scale modeling issues in LED, Proceedings Volume 7375, ICEM 2008: International Conference on Experimental Mechanics 2008; 737507 (2009) https://doi.org/10.1117/12.839005.

[23] G.Q. Zhang, Jinmin Li (eds.), Light-Emitting Diodes: Materials, Processes, Devices and Applications (Springer International Publishing, 2019).

[24] F.M. Steranka et al., High power LEDs–technology status and market applications, Physica Status Solidi (A), 194(2), 380–388(2002).

[25] S.W.R. Lee, R. Zhang, K. Chen, J.C.C. Lo, Emerging trend for LED wafer level packaging, Front. Optoelectron. 5(2), 119–126(2012), http://dx.doi.org/10.1007/s12 200–012–0259–9.

[26] Global LED Packaging Market Size study, By Power Range, By Packaging Material, By Application, and Regional Forecasts 2022–2028, May, 2022.

[27] E. Juntunen, et al., Copper-core MCPCB with thermal vias for high-power COB LED modules, IEEE Trans. Power Electron. 29(3), 1410–1417(2014), http://dx.doi.org/10.1109/TPEL.2013.2260769.

[28] W.B. Hardy, Developments in Polymer Photochemistry (vol. 3. Allen N.S., ed. London: Applied Science Publication; 1980). p. 322.

[29] Yoon Hwa Kim, Noolu S.M.Viswanath, Sanjith Unithrattil, Ha Jun Kim and Won Bin Im, Review—phosphor plates for high-power LED applications: challenges and opportunities toward perfect lighting, ECS Journal of Solid State Science and Technology, 7(1), R3134–R3147 (2018).

[30] M. Hamidnia, Y. Luo, X.D. Wang, Application of micro/nano technology forthermal management of high power LED packaging–a review, Appl. Therm. Eng. 145, 637–651(2018), http://dx.doi.org/10.1016/j.applthermaleng.2018.09.078.

[31] Linear Polydimethylsiloxanes, ECETOC Joint assessment of commodity chemical 63148–62–9 (1994).

[32] H.R. Fischer, C. Semprimoschnig, C. Mooney, et al. Degradation mechanism of silicone glues under UV irradiation and options for designing materials with increased stability. Polym Degrad Stab. 98, 720–726 (2013).

[33] Dow Corning Corporation. Silicone chemistry overview; 1997. www.dowcorning.com/content/ publishedlit/51–960A–01.pdf.

[34] D. Graiver, K.W. Farminer, R. Narayan. A review of the fate and effects of silicones in the environment. J Polym Environ. 11, 129–136 (2003).

[35] K. Xiang, G. Guangsu Huang, J. Zheng, et al. Accelerated thermal ageing studies of polydimethylsiloxane (PDMS) rubber. J Polym Res. 19, 1–7 (2012).

[36] Q. Zhang, J. Feng, C. Zhaohui, et al. Effect of temperature and moisture on the luminescence properties of silicone filled with YAG phosphor. J Semicond. 32, 2465–2473 (2011).

[37] Z. Yang, L. Feng, S. Diao, et al. Study on the synthesis and thermal degradation of silicone resin containing silphenylene units. ThermochemicaActa. 521, 170–175 (1970).

[38] A.A. Miller, E.J. Lawton, J.S. Balwit. Effect of chemical structure of vinyl polymers on crosslinking and degradation by radiation. J Polym Sci. 14, 503–504 (1954).

[39] R.B. Fox, L.G. Isaacs, S. Stokes. Electron spin resonance studies of photodegradation in poly(methylmethacrylate). J PolymSci A. 1, 1079–1083 (1963).

[40] M. Dole, The Radiation Chemistry of Macromolecules, vol. II.(London (NY): Academic Press) (1973), p. 97.

[41] J.A. Moore, J.O. Choi. Radiation effects on polymers. In: Clough R.L., Shalaby S.W., editors. Fundamentals of Space Systems. (New York (NY): Oxford University Press) 1991. p. 158.

[42] B. Ranby, J.F. Rabek. Photodegradation, Photooxidation and Photostabilization of Polymers, Principles and Applications. (New York (NY): John Wiley and Sons) 1975. p. 153.

[43] N. Grassie, M.I. Guy, N.H. Tennent. Degradation of epoxy polymers: part 1-products of thermal degradation of bisphenol-A diglycidyl ether. Poly Degrad Stab. 12, 65–91 (1985).

[44] P. Jain, V. Choudhary, I.K. Varma. Effect of structure on thermal behaviour of epoxy resins. EurPolym J. 39, 181–187 (2003).

[45] J. Macan, I. Brnardić, S. Orlić, et al., Thermal degradation of epoxy-silica organic-inorganic hybrid materials. PolymDegrad Stab. 91, 122–127 (2006).
[46] A. Gu, G. Liang, Thermal degradation behaviour and kinetic analysis of epoxy/montmorillonite nanocomposites. PolymDegrad Stab. 80, 383–391 (2003).
[47] Y. Liu, et al., Thermal behavior of flip chip LED packages using electrical conductive adhesive and soldering methods, 2013 10th China Int. Forum Solid State Light. China SSL (2013) 4–7, http://dx.doi.org/10.1109/SSLCHINA.2013.7177300, 3.
[48] S. Nakamura and G. Fasol, The Blue Laser Diode: GaN Based Light Emitters and Lasers (Springer) (1997).

5 Opportunities and Challenges in Flexible and Organic LED

Shalu C.

CONTENTS

5.1 INTRODUCTION

Due to recent advancements in the field of material science, flexible display devices experienced a lot of advantages. There are so many advantages to flexible display devices as compared to conventional ones as these are prone to any breakage as well as durable under even harsh operational conditions [1–5]. Flexible display devices provide the best suitability for printing applications and therefore the printing area can be bigger now [5,6]. Due to their suitability for printing, we can use these devices in various types of applications like wearable displays, electronic displays, hoardings, etc. Among all types of available displays, organic-based light-emitting displays are mostly used due to their advanced capabilities in terms of brightness, availability of different colors, power consumption, less temperature emission, and other parameters [7,8]. OLEDs have many advantages however they have some problems associated with their flexibilities that must be overcome so that we may use them for commercial purposes [9,10,11].

Presently, the main problems or challenges related to OLEDs, which are used for flexible purposes, can be summarized in the following ways: (a) instabilities in flexible substrate against temperature and chemical variations; (b) challenges with fixing and mechanical related stability on the plastic-based substrates, particularly in the case of electrodes made of metal or metal oxides; (c) trouble in patterning of deposited films caused by the lower solvent resistance offered by flexible foils and area needed, which is large. Here, in this review, the purpose is to highlight previously mentioned challenges in the fabrication of OLEDs. This review provides the reader a way to see what challenges have been resolved, the level of commercialization of flexible organic light-emitting devices, and the important references related to this context for the study.

5.2 THERMAL STABILITY OF TYPICAL FLEXIBLE SUBSTRATES

Various plastic-based foils (substrates) have been introduced to fulfill the need for optical as well as mechanical properties, for lightweight and flexible OLEDs [12]. We have substrates made of cellulose for the fabrication of organic light-emitting devices but those are not covered in this chapter as proper commercialization of these devices has not been done yet. Despite the fact that many of these foils have been tested for greater flexibility as well as transparency, these offer lower degradation temperature, which is around 100–150°C as compared to regular substrates. It is required for films to be deposited through a lesser temperature deposition process. Generally, foils like polyimide (PI) are not that transparent but show good thermal stability. It is required to have transparency greater than 95% in the visible range [13–16]. Substrates like polycarbonate, which is a transparent foil, exhibit modest solvent resistance.

Plastic substrates are required to have additional barriers as well as passivation films so that they may have protection against humidity and/or oxygen penetration and it is one of the most important challenges to overcome for the large level of commercialization of flexible organic light-emitting devices. But no such versatile and universal substrate has been identified, which can be used to fabricate flexible organic light-emitting devices however, polyethylene terephthalate (PET), as well as polyethylene naphthalate (PEN), are being used in the fabrication of such devices. These polymers are well suited due to their properties like suited thermal stability as well as solvent resistance and poor oxygen and moisture absorption. The flexing ability of polyethylene napthalate (PEN) substrate along with fabricated pixels on its surface, as shown in Figure 5.1.

Though plastic-based material provides greater thermal stability however they are difficult to bend after some time and hence, due to increased bending curvature,

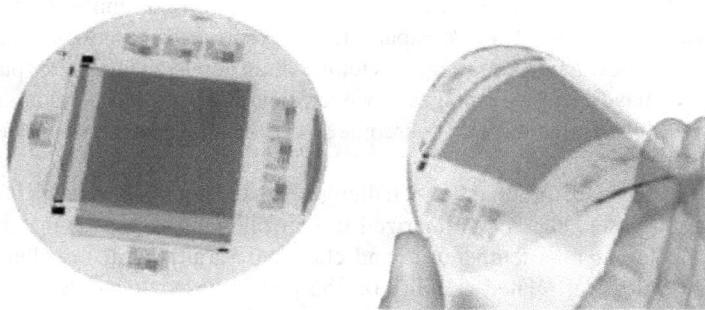

FIGURE 5.1   Pixel array fabricated over the flexible PEN plastic substrate [10]. Copyright © 2016 Mariya Aleksandrova.

there are restrictions on the rolling of the devices. As the fabrication process is carried out at higher temperature than 100°C, the biggest challenge in the fabrication of such devices with plastic material is that those experience thermal expansion as well as degradation at such a temperature even though they are having very good properties including their lower cost [13, 17–19]. Such degradations may be understood as in the case of any foil subjected to vacuum sputtering of ITO on PET, it experiences residual stress. At the time of film growth, this is the required modification of deposition modes that may lead to the amorphous microstructure of the oxide films. These amorphous films provide increased sheet resistance as compared to the crystalline phase and hence we experience increased turn-on voltage of organic light-emitting devices. While working continuously in extreme conditions like maximum brightness (respectively, maximum current density) there is always a chance for foils to face thermal degradation. As a result, the sheet resistance of conductive film, which is deposited on polyethylene naphthalate substrate increases by 20%. It has been demonstrated in various references that PET foil at around 115°C deforms and its bending radius of curvature increases to nearly 1 meter. As far as tensile stress and compressive stress are concerned, polyethylene terephthalate (PET), as well as polyethylene naphthalate (PEN), show the same response at room temperature because these seem stable even with bending greater than 20000 also, they do not crack. Polyethylene terephthalate (PET), as well as polyethylene naphthalate (PEN), show chemical stability as they are resistant not only to cleaning solutions like isopropanol and ethanol but also to organic solvents like chlorobenzene, toluene, and chloroform. It has been observed that glycol modifications of polyethylene terephthalate, show considerable resistance to acetone when treated for a longer duration. Due to the hydrophobic properties of flexible substrates, any plastic surface needs to exhibit wetting properties so that it may have organic solutions [20]. Many pretreatments have been established, which are used before the deposition process so that surface tension may decrease and may increase the wetting conditions. We use various methods in industries like the low-pressure cold gas plasma method, corona discharge, and flame plasma UV zone treatment.

The method which is to be used depends on the material used. We can understand it with an example. We know that polyethylene terephthalate is having a low melting point (nearly 100°C) and this can be damaged if we use it with flame plasma because the heat generated in this process will damage it. However, we can use either polystyrene, which is having a melting point of around 200°C, or polyimide, which is having a melting point of 270°C.

5.3 MECHANICAL STABILITY OF OLED STRUCTURES UPON ROLLING AND BENDING

One of the biggest challenges in the commercialization of organic light-emitting devices is that there is a deficiency of very flexible electrodes, which are transparent. We use indium tin oxide in the manufacturing of glass-based organic light-emitting devices however it exhibits poor mechanical flexibility. Other suitable transparent electrodes are also in the process of development, which will exhibit flexible properties. However, it is important in the development that there should be a proper combination of transparency as well as sheet resistance so that it may fulfill the requirements of the manufacturing of organic light-emitting devices. Intrinsic stress developed in a thin film due to curvature provides an indication of whether the coating is having high mechanical stability or having a lower one. On the other hand, there are various factors that affect intrinsic stress as there exists a stress gradient along the radius of curvature. Also, there exist some other errors because of nonlinear effects that include areas and shape. So, how sheet resistance changes and microcracks in a given area at dynamic or static bend mode are preferred to evaluate the criteria for flexible conductive films. As per various investigated multilayer systems, the direction of the cracks is generally 90 degrees tilted towards the mechanical loading, which does not depend on inner or outer bending. To inspect surfaces generally high-resolution transmission electron microscopy, scanning electron microscopy, and atom force microscopy are used. The four-point probe measurement technique is used for the accurate measurement of the change in sheet resistance. Another studied substitute for indium tin oxide on flexible substrate is the indium tin oxide-free electrode, which is transparent in nature and has a multilayer system of metal oxide/metal/metal oxide.

Generally, metal oxide/metal/metal oxide can be replaced with other transparent oxides having nonconductive properties. Ductile metals like silver, which provide conductivity, are used in between layers. The thickness of the metal film is kept much less (<20 nanometer) so that in the visible spectrum, it may decrease overall transmittance. There are references and studies available for the electro-optical and mechanical characteristics of $ZnO/Ag/ZnO$, $Al_2O_3/Ag/Al_2O_3$, ultra-thin silver film, inserted between Mn-doped tin oxide films, $MoO_3/Ag/MoO_3$, $TiO_2/Ag/TiO_2$, and so on.

Here there are some contradictions in the studies as some authors claimed that devices made of these materials show very high optical transparency nearly 90% for visible range but they could not mention anything about variation in resistances offered by a film under stress. Many authors considered a smaller radius of curvature in order of 6mm and a repeating bend cycle of 5000 that the overall structure may withstand with a very small change in sheet resistance. Well, the best device fabricated so far could provide average optical transparency lower than 85%. Mechanical test specifications are not in detail in the case when there is an optimum tradeoff between conductivity, mechanical stability, and transparency.

Generally, in the study, the bending cycle is considered as 1 Hz whereas the radius of curvature has been considered as 10 mm. There is no detailing about reliability over a while however details regarding turn-on voltage, luminescence, current, efficiency and other specifications of optoelectronic structure.

There are very good results for sulfite-based electrodes like $ZnS/Ag/WO_3$. The change in sheet resistance depends on the radius of curvature and it results in the tensile stress among the layers and is nearly unchanged for a radius <5 mm (Figure 5.2). When mechanical flexibility is compared with the indium tin oxide-based electrodes on polyethylene terephthalate flexible substrate, the first one is very poor. At a radius of curvature of 10 mm and 50 cyclic bends, the change in sheet resistance is around 49%.

In addition, the contact between $ZnS/Ag/WO_3$ electrode and the organic films, NPB tends to a reduction in the barrier potential for the injection of holes, which

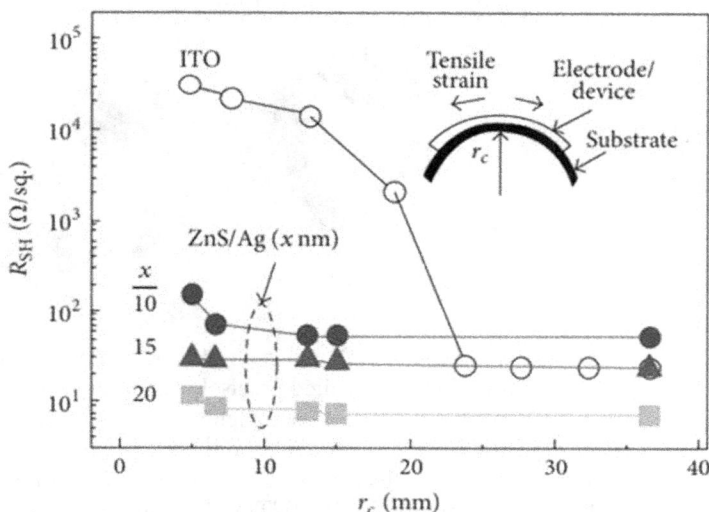

FIGURE 5.2 Sheet resistance variation of ZnS/Ag/WO$_3$ films on PET with a radius of curvature after bending [10]. Copyright © 2016 Mariya Aleksandrova.

in turn helps in achieving high current efficiency. Graphene is one of the most promising candidates for the fabrication of flexible OLEDs as it may be used as a transparent conductive electrode owing to its good optical transmittance and conductivity. Additionally, it has great mechanical strength, flexibility, and chemical stability [21].

The growth of graphene on large Cu foil can be achieved by using low carbon solubility in Cu23. On the other hand, poly methyl-methacrylate (PMMA) can also be used to transfer the graphene over the flexible PET sheets with a smaller number of defects. Up to 3mm radius of bend cycle (tensile strain ~5%), the resistance in lateral and transversal directions remains approximately independent. When the radius of the bend cycle is reduced to 1 mm, the transversal direction resistance is reduced by a factor of 2. So, a new approach to getting a low sheet resistance gains attention, which is the growth of graphene on a large scale by the chemical vapor deposition (CVD) method by using a nonvolatile ferroelectric polymer. The graphene sheets doped by poly(vinylidene fluoride-co-trifluoroethylene) (PVDF-TrFE), yield a low sheet resistance of approximately 120Ω/square with 95%transmittance in the visible range. This is achieved mainly due to the highly transparent nature of PVDF. The film folding capacity on the PET substrate is measured by estimating the variation of its resistance with the bending radius, is shown in Figure 5.2. It is evident from Figure 5.2 that resistance is increased by a small amount at a radius of bending approximately 3 mm, and the initial value of resistance is restored several times even after bending of the sample to radius post 1mm. This improved performance in terms of mechanical properties of all structures in comparison to the ITO-based devices is attributed to the intermediate metal films with good ductility. Even though these structures are complex having more than at least three layers. Additionally, they require an

FIGURE 5.3   Roll-to-roll deposition of organic films on foil.

expensive vacuum deposition process, which may restrict their size. Additionally, the thickness of metal film should also be controlled precisely because elastic properties are strongly dependent on the thickness on a nanometer scale. In recent times, the inorganic components in the LEDs have been replaced by organic materials to make flexible OLEDs. Several transparent electrodes such as conducting polymers [22], graphene, multi-wall carbon nanotubes (MWCNTs) [15], and single-wall carbon nanotube (SWCNT) films have become suitable alternatives in the field of flexible OLEDs. The sheet resistance of these electrodes is limited owing to their hopping conduction mechanism. Some of the structures demonstrate a variation in the sheet resistance from 8% and 16% with the bending radius of 9and 6mm, respectively. The repeating bend cycles are greater than 300 for this measurement, which are generally larger in comparison to the metal oxide/metal/metaloxide electrodes. In addition, all the fabrication steps are easier and more cost-effective. To improve electrical conductivity further an intermediate metal film with an appropriate refractive index between two polymeric films is inserted. It was observed that the electrodes poly(N-vinylcarbazole)(PVK)/Ag/PVK and PVK/Ag/PEDOT:PSS demonstrated favorable electro-optical properties. For example, the sheet resistance maybe reduced down to 10Ω/square with the optimum transmittance larger than 85%. Additionally, the sheet resistance of less than 2% after 500 bends has been reported, which fulfills all the current necessities of the consumer flexible optoelectronics.

5.4   ORGANIC EMISSIVE AND CHARGE CARRIERS TRANSPORTING FILM DEPOSITION AND PATTERNING OVER FLEXIBLE SUBSTRATES

Even though there are no serious issues related to the mechanical stability of colored organic films in the flexible OLEDs, there are a few serious shortcomings associated with the patterning for pixel development. Traditional coating methods for patterning are photolithography and lift-off. The etching and stripping solutions cannot be applied to these substrates because they have low solvents' resistance. This requires the advancement of new preservative deposition methods, which provides direct stamps. Additionally, it fulfills all the requirements of large area coatings in a cost-effective and high-quality in an effective manner [5,22]. Recently, roll-to-roll (R2R) printing of polymer films on PET sheets became a very common manufacturing process. The main concept is to deposit the organic inks on a pre-defined site, which forms the pattern directly on rollable substrates in one step only [8,21]. The key advantage of printing approaches is its prospect of direct patterning while deposition irrespective of contact or noncontact methods.

In this manner, the protection methods for the flexible substrates are required and post-deposition treatments typically for the photolithography solvents are circumvented. In screen printing, the required pattern is primarily described by the openings in the screen created by photolithography. In addition, the density of the openings (mesh) defines the particle size of the organic substance, which can cross the openings when a wiper crosses over the ink. An additional binder polymer should be mixed with the functional organic compounds to give the required thixotropy of the inks in addition to their viscosity. Generally, the thickness of the active layer in the organic devices lies in the range from 100 nm to 50 μm [13–15]. It has been observed that the heat dissipation in a thick screen-printed film, which lies generally in the range of μm is better due to the larger volume of the functional organic material. Figure 5.4 shows the variation of luminous efficiency of a screen-printed layer of low-molecular-weight compound OLED with polymeric binder at different currents and after 400 repeating bends. The average value of luminous efficiency is 3.5 cd/A and the power consumption in this current range is very less (from 35.7mW to 73.4mW). The microscopic images of the screen-printed organic electroluminescent layer with a typical thickness of 2 μm are shown in Figure 5.4(b). It is evident from the image that the distribution of organic particles is continuous throughout the film without microcracks, which suggests the mechanical stability of the thick films in the presence of binder substance. There is only a redistribution of the larger clusters on the surface at a bending radius of 5 mm, probably due to remaining unhomogenized particles during ink preparation. On the other hand, the heat dissipation is not sufficient in the

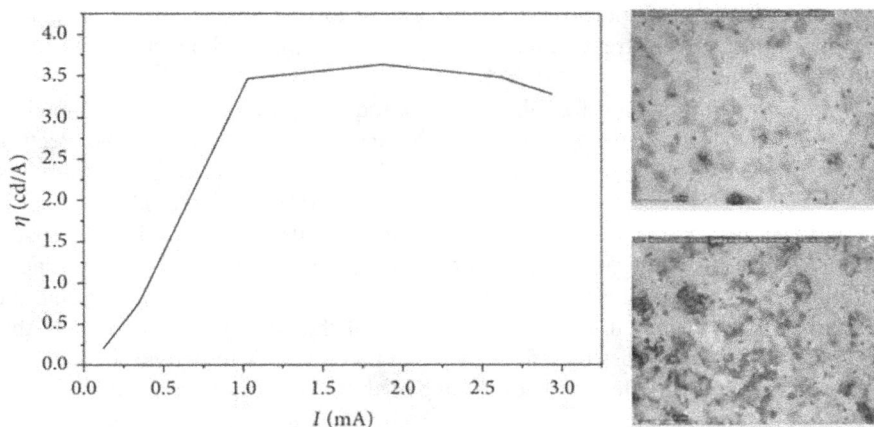

FIGURE 5.4   (a) Screen-printed layer OLED luminous efficiency with low-molecular-weight compound and polymer binder after 400 bends. (b) Microscopic images of the screen-printed organic electroluminescent layer before (up) and after bending (down) [10] Copyright © 2016 Mariya Aleksandrova.

thin screen-printed ink heat. So, the heat is gathered in the structure, which may cause the melting of the flexible substrate at the current values larger than a few mA/cm^2. The main limitation in this process is the modest resolution of the stamp because the mesh dimension cannot be very low beyond a certain limit. The stamp is generally made of sufficiently large organic particles with a size comparable to the openings of the mesh. For ink reflowing, thermal treatment of the screen-printed layers is essential in addition to homogenization and stabilization of the imprint. The annealing temperature for thermal treatment should be lower than the deformation temperature of the flexible foil. However, it may limit the choice of inks and binders. The residual solvent or underflowed clusters may enhance the roughness of the surface of an organic layer.

Enhancement in the roughness causes the contact resistance increment and eventually hampering of luminous efficiency, typically < 5 cd/A. Further, a rotary printer with a moving substrate between a cylindrical mask and impression cylinder technique for screen printing can also be employed to increase the throughput. But unfortunately, this technique is still under test for organic materials on flexible substrates such as foil.

Another method namely inkjet printing utilizes a nozzle head to place low viscosity inks. In this method, the droplets, which may be ejected can be very fine due to the ultrasonic generation of the aerosol flow as in electrospray systems [8]. Additionally, the thin films upto a few nanometers can be simply customized and the volume control typically in the range of picoliters is attainable at the head of the printer head. A few film defects such as cracking, intrinsic stress, and peeling off may appear during the drying of the films due to limitations of solvent type and volatility of the solvent. To increase the density of molecules in the films the rate of evaporation of solvent should be maintained. The next limitation of this method is to achieve a uniform distribution of the droplet over these flexible substrate surfaces. For this purpose, prior treatments of these plastic surfaces such as UV, plasma, or solvent treatments are employed to improve the wetting ability of printed patterns [8,17].

Another popular method for OLED fabrication is Gravure printing. This method is also employed for the deposition of a hole transporting and emissive and electron transporting layers on the flexible substrate. This technique is having faster production rate than that of roll-to-roll printing processes [15]. In this method, the printing roll and picking ink from the reservoir are inscribed cells for the printing of the surface, as shown in Figure 5.3. The form of cell with the imprint density is variable. Additionally, a special blade is used to remove the excess ink and to distribute the ink uniformly in each cell. The main limitation of the OLED fabricated using this method is the limited emission efficiency. Even though a few researches have shown encouraging and favorable outcomes, Chung et al. reported high-performance PLED with a maximum quantum efficiency of 8.8 lm/W and maximum luminance of 66000 cd/m^2 [12].

Furthermore, techniques such as flexography printing [13], microcontact printing, and nanoimprinting [14] are anticipated to receive more popularity in the future and required optimizations in the process because many hurdles hamper their industrial use.

5.5 CONCLUSIONS

All organic, i.e., flexible OLED is one of the best alternatives for the upcoming display technologies owing to their benefits such as portability with large areas and inexpensive fabrication. Despite several advantages still, many challenges should be taken care of in the fabrication of flexible OLEDs. When plastic substrate sare used to fabricate the OLED fabrication, the device shows several superior characteristics over the glass substrate-based devices, such as being lightweight, cost-effective, and flexible. The reliability and lifetime of these devices are highly related to thermal expansion and moisture absorption plastic materials. The electro-optical properties of electrode coatings govern their mechanical properties and cracking behavior. Generally, these characteristics are highly dependent on deposition parameters, the method of doping, and the morphology of the film. In this chapter, a review of some of these characteristics has been performed. Mostly, all the fabrication methods are not useful for large-area fabrication of the devices and have certain limitations, such as speed, defects in the coatings, and resolution. The roll-to-roll line fabrication is required to overcome this limitation of speed and large-area fabrication. In addition, the use of alternative less harmful solvents for processing the conjugated polymers should also be employed, which will be further useful in the dry printing development.

ACKNOWLEDGMENT

The author is grateful to Prof. Intezar Mahdi, Director, SET, IFTMU for his continuous encouragement and support for research.

REFERENCES

1. C. Cho, K.L. Wallace, P. Tzeng, J.-H. Hsu, C. Yu, and J.C. Grunlan (2016). Outstanding low-temperature thermoelectric power factor from completely organic thin films enabled by multidimensional conjugated nanomaterials. Adv. Ener. Mater. 6 (7), 1502168.
2. C. Shalu, N. Yadav, K. Bhargava, M.P. Joshi, and V. Singh (2018).All organic near-ultraviolet photodetectors based on bulk hetero-junction of P3HT and DH6TSemiconductor Science and Technology33 (9), 095021.
3. M.-C. Choi, Y. Kim, and C.-S. Ha (2008).Polymers for flexible displays: from material selection to device applications. Progress in Polymer Science, 33 (6), 581–630.
4. L. Gao, J. Song, J.U. Surjadi, K. Cao, Y. Han, D. Sun, et al. (2018). Graphenebridged multifunctional flexible fiber supercapacitor with high energy density. ACS Appl. Mater. Interfaces 10 (34), 28597–28607.
5. S. Chaudhary, K. Bhargava, N. Yadav, M.P. Joshi, and V. Singh (2019). Effect of concentration of DH6T on the performance of photoconductor fabricated using blends of P3HT and DH6T, Optical Materials 89, 214–223.
6. G. Gustafsson, Y. Cao, G.M. Treacy, F. Klavetter, N. Colaneri, and A.J. Heeger (1992). Flexible light-emitting diodes made from soluble conducting polymers. Nature 357 (6378), 477–479.
7. W. Bock (2005). Advances in flexible electronics displays, Pira International.
8. C. Shalu, S.R. Mohan, M.P. Joshi, and V. Singh. Structural and optoelectronic characterization of organic vapor phase deposited thin films of oriented DH6T molecules, American Institute of Physics Conference Series 1832 (8).

9. X. Feng, M.A. Meitl, A.M. Bowen, Y. Huang, R.G. Nuzzo, and J.A. Rogers (2007). Competing fracture in kinetically controlled transfer printing. Langmuir 23 (25), 12555–12560.

10. M. Aleksandrova (2016). Advances in Materials Science and Engineering, 4081697.

11. D. Corzo, G. Tostado-Blázquez, and D. Baran (2020). Flexible electronics: status, challenges and opportunities, Frontiers in Electronics, 1, 594003.

12. J.W. Park, D.C. Shin, and S.H. Park (2005). Large-area OLED lightings and their applications, Semiconductor Science and Technology, 26 (3),034002.

13. K.J. Allen (2005). Reel to real: prospects for flexible displays, Proceedings of the IEEE, 93 (8), 1394–1399.

14. J.H. Kim, D.-H. Kim, and T.-Y. Seong (2015). Realization of highly transparent and low resistance $TiO_2/Ag/TiO_2$ conducting electrode for optoelectronic devices, Ceramics International, 41(2),3064–3068.

15. W.S. Wong and A. Salleo, Flexible Electronics: Materials and Applications, Springer Science & Business Media, Berlin, Germany, 2009.

16. P.E. Burrows, G.L. Graff, M.E. Gross et al. (2001). Ultra barrier flexible substrates for flat panel displays, Displays, 22 (2), 65–69.

17. S. Ummartyotin, J. Juntaro, M. Sain, and H. Manuspiya (2012). Development of transparent bacterial cellulose nanocomposite film as substrate for flexible organic light emitting diode (OLED) display, Industrial Crops and Products, 35(1), 92–97.

18. S. Khan, L. Lorenzelli, and R.S. Dahiya (2015). Technologies for printing sensors and electronics over large flexible substrates: a review, IEEE Sensors Journal, 15(6), 3164–3185.

19. M. Strobel, M.J. Walzak, J.M. Hill, A. Lin, E. Karbashewski, and C. S. Lyons (1995). Comparison of gas-phase methods of modifying polymer surfaces, Journal of Adhesion Science and Technology, 9(3),365–383.

20. J.-A. Jeong and H.-K. Kim (2013). Al2O3/Ag/Al2O3 multilayer thin film passivation prepared by plasma damage-free linear facing target sputtering for organic light emitting diodes, Thin Solid Films, 547, 63–67.

21. K.S. Kim, Y. Zhao, H. Jang et al. (2009). Large-scale pattern growth of graphene films for stretchable transparent electrodes, Nature, 457(7230), 706–710.

22. S. Yadav, V. Kumar, S. Arora, S. Singh, D. Bhatnagar, and I. Kaur (2015). Fabrication of ultrathin, free-standing, transparent and conductive graphene/multiwalled carbon nanotube film with superior optoelectronic properties, Thin Solid Films, 595, 193–199.

6 Light Extraction Efficiency Improvement Techniques in Light-Emitting Diodes

*M. Manikandan, G. Dhivyasri, D. Nirmal,
Joseph Anthony Prathap and
Binola K. Jebalin I. V.*

CONTENTS

6.1 INTRODUCTION

Gallium nitride (GaN)-based optoelectronic devices have significantly impacted solid-state lighting. During the past several decades, III-nitride semiconductor materials have emerged as the leading material for light-emitting diodes (LEDs) with an emission spectrum spanning from ultraviolet, blue, and green wavelengths. This enables large-scale fabrication of full-colored LED displays and white LEDs. In such cases, it requires blue LEDs with high-efficiency [1]. But, high-efficiency LEDs emitting in blue regions are still confronting the effect of high injection current efficiency droop. The researchers have not yet identified an exact solution to the efficiency droop problem. However, enhancing LED performance, investigating unavoidable

DOI: 10.1201/9781003340577-6

challenges in solid-state LED, and modeling quantum phenomena still have space for pushing the LED to meet future demands [2]. Although many explanations were proposed, the physical origin of efficiency droop is not well understood. So far, many techniques have evolved to tackle the efficiency droop effect in III-nitride-based LEDs. Based on this, numerous mechanisms are proposed to account for efficiency droop; quenching current injection efficiency, Auger recombination, carrier delocalization, and carrier leakage are the main emphasis of the researchers. In addition to these mechanisms, the lack of high-quality substrate material in the epitaxial growth of nitride material recedes advancement in nitride LEDs. Commercial nitride-based LEDs are currently heteroepitaxial grown on sapphire due to their durability and low cost. The larger lattice mismatch (13.0%) of sapphire with GaN epitaxial layer induces huge dislocation density and reduces the crystal quality in the epitaxial structures. In turn, diminishing the life period of LEDs and device performance. The effective substitute material for sapphire is silicon carbide substrate. The lattice mismatch lesser than 3.40% with the GaN epitaxial layer is appropriate for GaN LED fabrications.

In this chapter, a few techniques were discussed, which enhances the optical performance in III-nitride LED.

6.2   EFFICACY OF LIGHT EMITTING DIODES

The GaInN/GaN LEDs' light output power is well-defined as the number of photons produced per unit of time and volume, which determines the brightness of LEDs. The light output power of LEDs is proportionate to their external quantum efficiency. The number of photons emitted from the device to the number of injected electrons in external quantum efficiency can be further decomposed as

$$EQE = IQE \times LEE \times \eta_{INJ} \qquad (6.1)$$

where
LEE – light extraction efficiency
IQE – internal quantum efficiency
η_{INJ} – injection efficiency

The light extraction efficiency does not have any role in GaInN/GaN LEDs' efficiency droop but it is related to device geometries. Thus, light extraction efficiency is the ratio of the number of photons generated in the LED to the number of photons emitted from the LED [3]. The ratio of the number of electrons supplied by the power sources to the number of electrons injected into the LEDs is termed as injection efficiency. Generally, injection efficiency is not a factor of GaInN/GaN LEDs assumed to be 100% of efficiency droop. The internal quantum efficiency is the ratio of the number of electrons injected into the LED to the number of photons produced inside the LED. In some devices, injection efficiency and LEE are constant and also EQE is proportional to IQE. To improve and analyze the efficiency droop of GaN/GaInN LEDs, IQEs play a vital role. The IQE of LEDs is categorized by the ABC model known as the carrier rate equation model with ABC coefficients [4].

6.3 LIGHT EMITTING DIODES' EFFICIENCY DROOP

The efficiency droop effect in LEDs at high injection current is one of the major research threads in the field of optical device design. Based on these many techniques are presented [5], such as tapered AlGaN EBL [6], the use of a multi-layer barrier in InGaN/GaN/InGaN [7], and quantum barrier in GaN/AlGaN/GaN [8]. The presence of the AlGaN/InGaN barrier is proposed in the MQW structures [9, 10].

The major challenges in LEDs are light extraction efficiency and internal quantum efficiency [11] and the key factors of LEDs' efficiency droop are quantum confined Stark effect, junction heating, Auger recombination, poor hole injection efficiency, current crowding, polarization effects. Electron blocking layer, staggered quantum wells, thin last barrier, hole reservoir layer techniques are adopted to decrease the efficiency droop [12, 13].

The radiative recombination rate $(B.n^2)$, auger recombination rate $(C.n^3)$, radiation efficiency in the quantum well (η radiative), non-radiative recombination rate $(A.n)$, auger efficiency (η_{Auger}) and non-radiative efficiency in the quantum well ($\eta_{non\text{-}radiative}$) are specified in Eqns. (6.2), (6.3), and (6.4) [14–18].

$$\eta_{radiative} = \frac{B.n^2}{A.n + B.n + C.n} \tag{6.2}$$

$$\eta_{non-radiative} = \frac{A.n}{A.n + B.n^2 + C.n^3} \tag{6.3}$$

$$\eta_{Auger} = \frac{C.n^3}{A.n + B.n^2 + C.n^3} \tag{6.4}$$

Representation of IQE is [19]

$$IQE = \frac{B.n^2}{A.n + B.n^2 + C.n^3 + k(n - n_0)^m + \dfrac{I_{LK}}{qV_{QW}}} \tag{6.5}$$

where
$k(n - n_0)^m$ – non-radiative carrier loss owing to carrier delocalization
n – quantum wells' carrier density
V_{QW} – active region volume
I_{LK} – carrier leakage current

6.4 LIGHT EXTRACTION EFFICIENCY (LEE) IMPROVEMENT TECHNIQUES

The LEE is well-defined as optical power emitted into free space. A few light extraction efficiency enhancement techniques are discussed below.

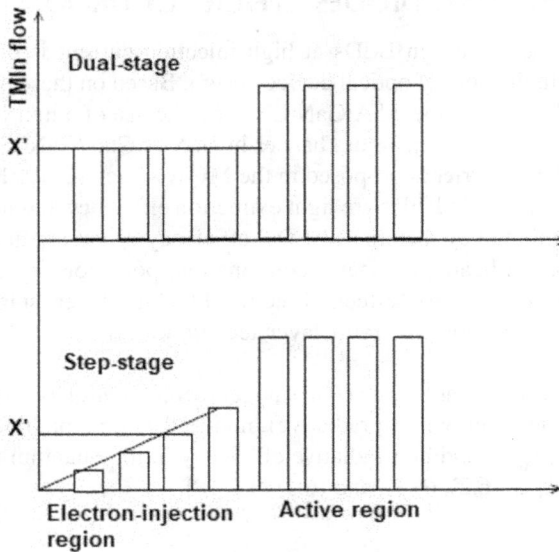

FIGURE 6.1   Step-stage vs. dual stage MQW LEDs.

6.4.1   Step Stage GaN/InGaN Multi-Quantum Well Structure

H. C. Hsu verified the step-stage MQW blue LEDs experimentally, where stepwise and constant indium-doped MQW structures are grown by monitoring the growth of the electron injection layer in TMIn flow rate. Step-stage LED technique enhances external quantum efficiency and optical output power compared to dual-stage LED due to the active minimization of the step-stage MQW blue LEDs' piezoelectric field [20]. Figure 6.1 represents the disparity of light output power as a function of current for dual-stage and step-stage MQW blue LEDs. From Figure 6.2, this step-stage LED offers high light output power related to that of dual-stage MQW blue LED.

6.4.2   Micro-Cavity OLED with Diffusion Layer

Cheol Hwee Park et al. [21] have investigated the impact of diffusion layer in MC-OLED as shown in Figure 6.3, which is composed of a transparent polymer poly methyl methacrylate with nano-sized structure combined with semi-planarization layer of zinc oxide consisting of $n = 2.10$ high refractive index in the device to enhance the light extractions and view angle characteristics efficiency.

Materials specifically, indium zinc oxide and silver are utilized as an anode, N'-bis (naphthalene-1-yl)-N, N'-bis (phenyl) benzidine as a hole transport layer, 1,4,5,8,9,11 hexaazatriphenylene hexacarbonyl nitrile and N, as the lithium fluoride, electron transport and emitting layer as the electron injection layer and aluminum as the cathode. Figures 6.4, 6.5, and 6.6 prove that micro-cavity (MC) LEDs have enhanced characteristics in terms of both electrical and optical parameters.

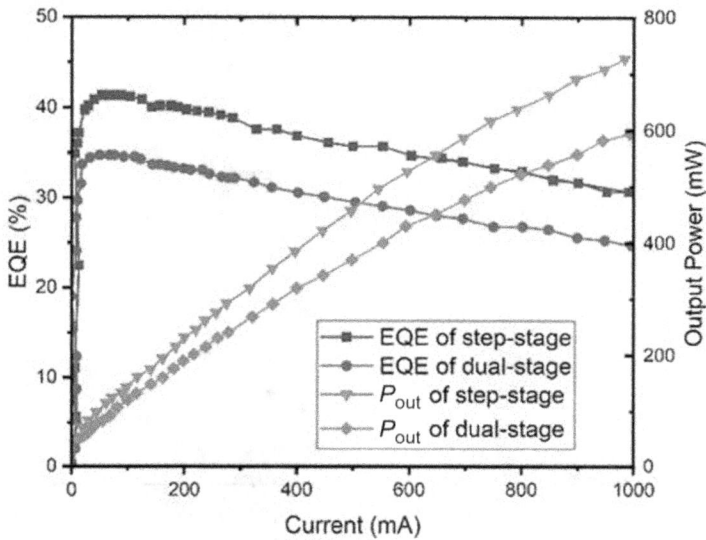

FIGURE 6.2 Comparison of external quantum efficiency, output power with respect to current.

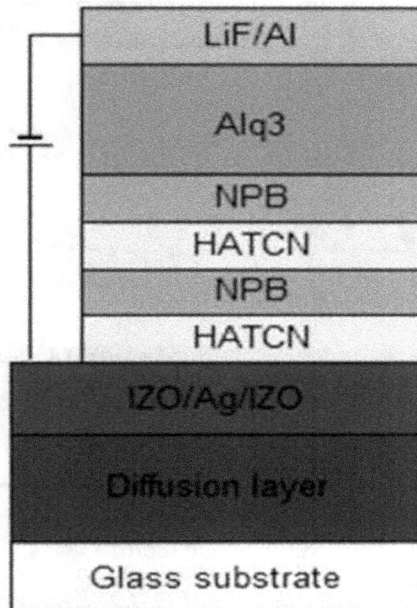

FIGURE 6.3 Schematic of MC-OLED structure with diffusion layer.

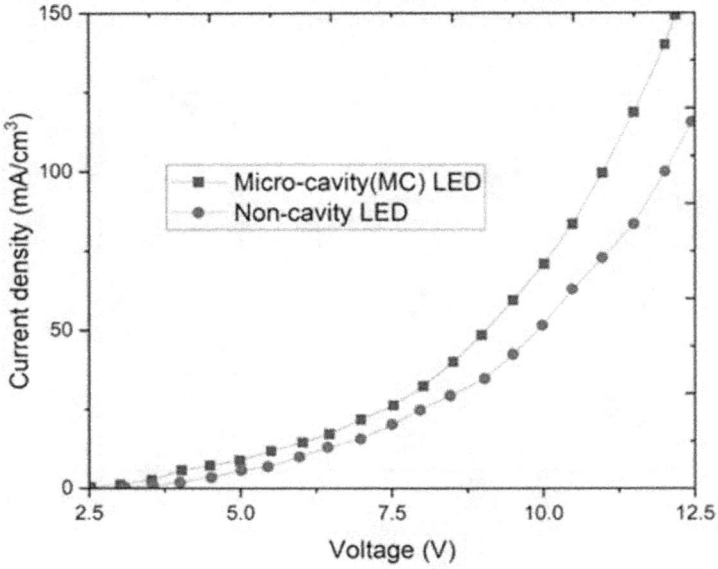

FIGURE 6.4 *V-I* characteristics for micro-cavity and non-cavity LEDs.

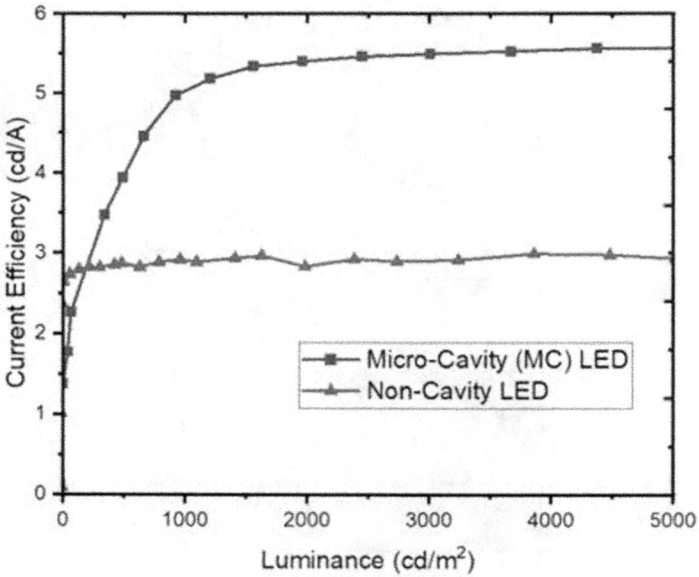

FIGURE 6.5 Luminance vs. current efficiency for micro-cavity and non-cavity LEDs.

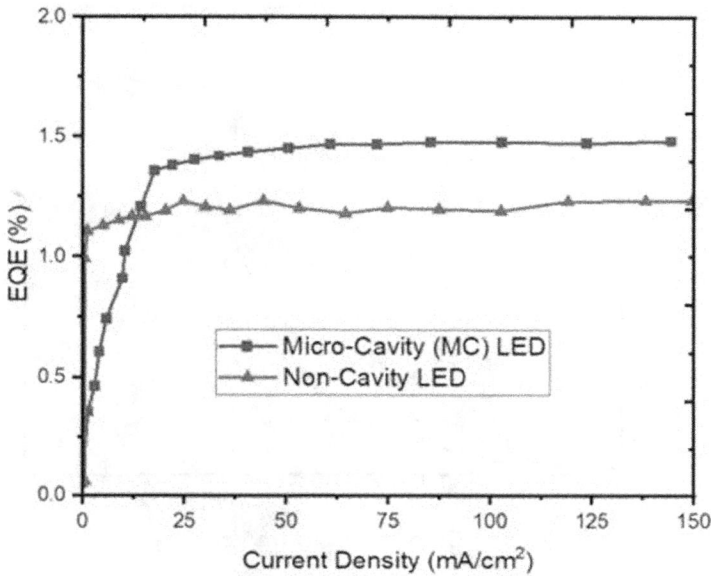

FIGURE 6.6 EQE vs. current efficiency for micro-cavity and non-cavity LEDs.

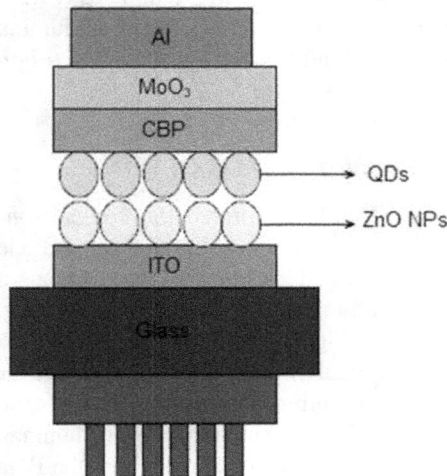

FIGURE 6.7 Schematic of ZnO nanopillar arrays in quantum dot LEDs.

6.4.3 ZnO Nanopillar Arrays in Quantum Dot Light-Emitting Diodes

In 2014, X. Yang et al. [22] have reported a QLED with layers of large-scale, periodic zinc oxide nanopillar array shown in Figure 6.7, which provides an enhanced light out-coupling efficiency. The uniform zinc oxide nanopillar arrays are attained by an

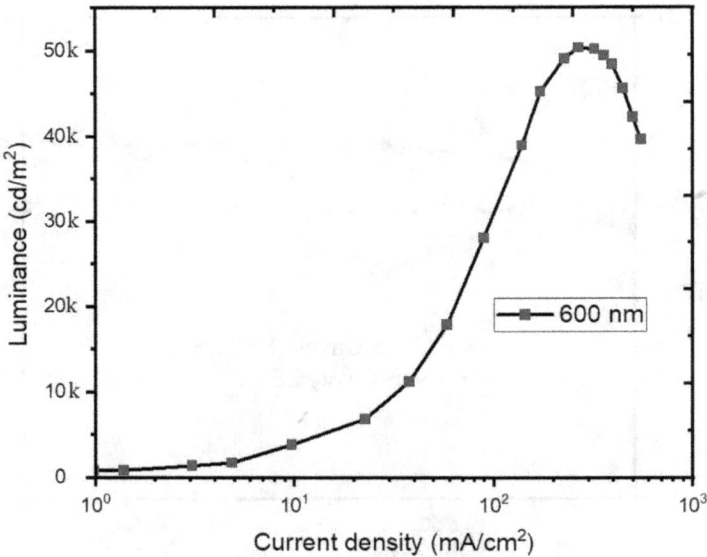

FIGURE 6.8 Current density vs. luminance of quantum dot LEDs.

efficient pattern repetition in the non-wetting templates (PRINT) techniques, which is capable of providing the current efficiency, maximum luminance, and EQE values of 54200 cd/m², 26.60 cd A⁻¹, and 9.341%, respectively (Figure 6.8, 6.9, and 6.10) compared with flat QLEDs.

6.4.4 BGaN Quantum Well-Based GaN LED

In 2022, Manikandan et al. [23] performed the comparison between the optical performances of GaN well and BGaN well light-emitting diodes. The anode voltage versus luminous power is illustrated in Figure 6.11. The comparative study results that luminous power increase with anode voltage, where else the radiative rate increase with anode voltage is shown in Figure 6.12, irrespective to the GaN well-LED devices. Thus, luminous power of the BGaN well light-emitting device is observed to be higher.

The investigation of injection current versus luminous power are observed to be a significant factor in radiative rate. Thus, injection current and luminous power are examined and presented in Figure 6.13. From the obtained graph, the BGaN well light-emitting diodes' luminous power is relatively higher than the GaN well light-emitting diode. These results are due to the higher radiative rate in BGaN well light-emitting diodes.

6.4.5 Compositional Step Graded InGaN Barrier Multiple Quantum Wells Light-Emitting Diode

Prajoon et al. [24] investigated a CSG InGaN barrier LED chip size of 300 μm × 300 μm depicted in Figure 6.14, which describes the substantial enhancement in internal

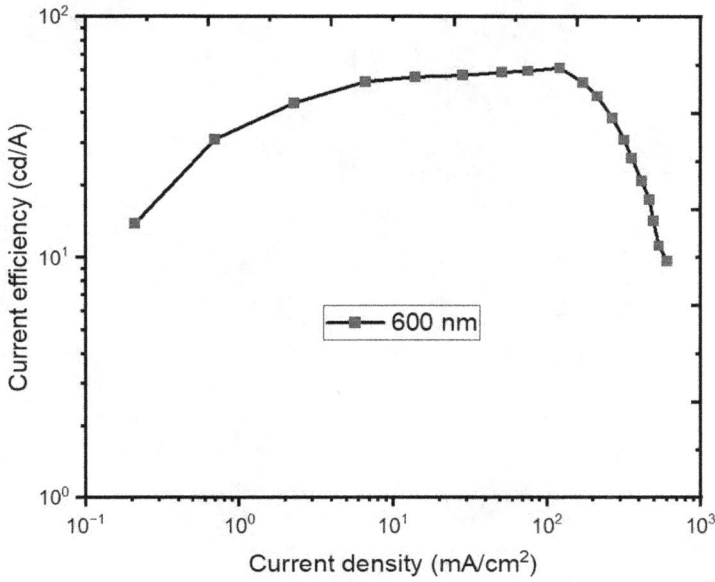

FIGURE 6.9   Current density vs. current efficiency of quantum dot LEDs.

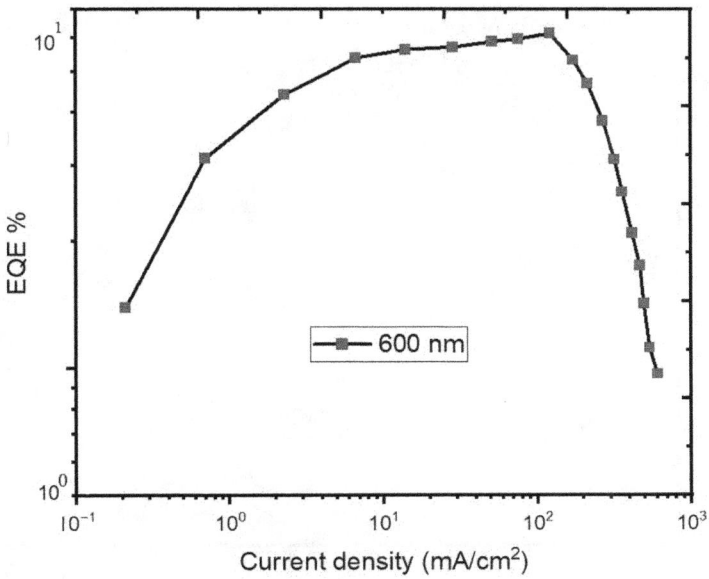

FIGURE 6.10   Current density vs. EQE of quantum dot LEDs.

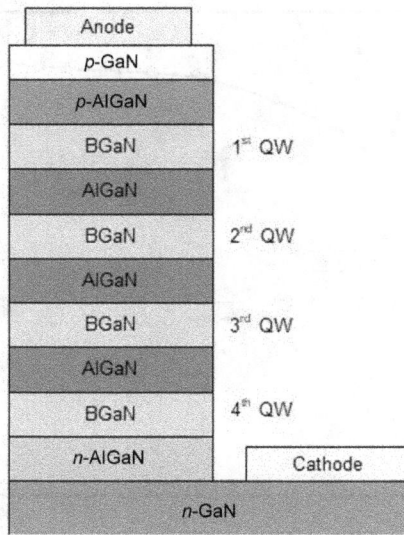

FIGURE 6.11   Schematic structure of BGaN well-based LED.

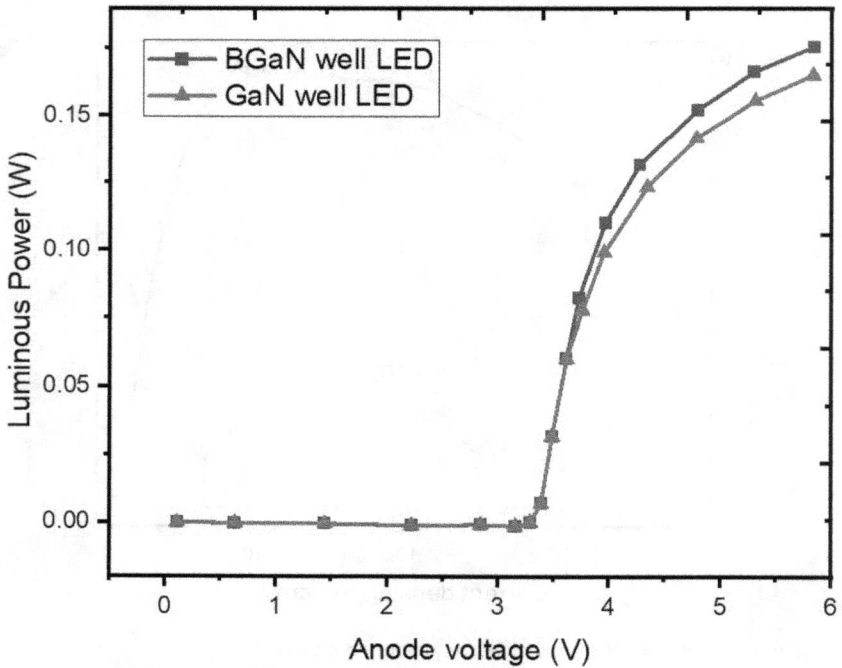

FIGURE 6.12   Luminance power vs. anode voltage for BGaN and GaN well LED structures.

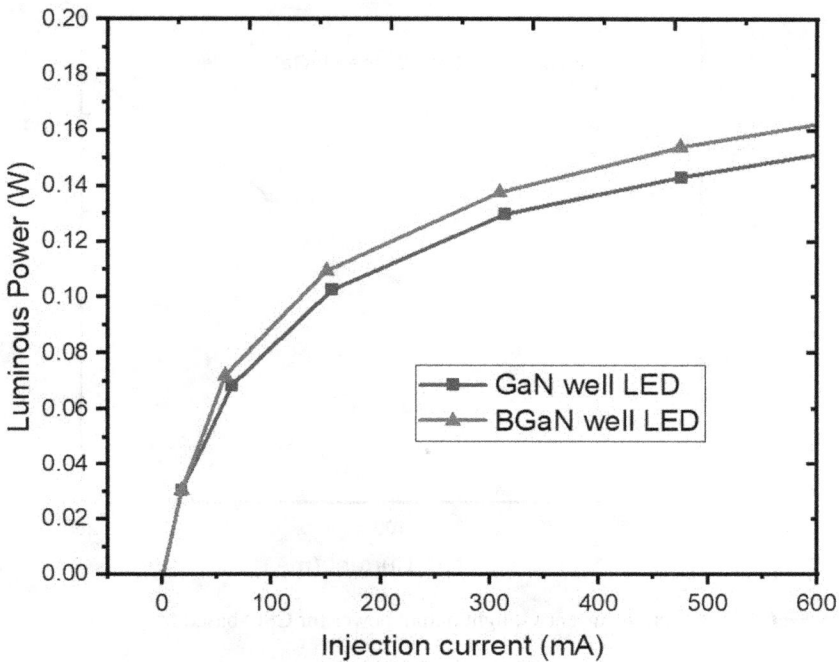

FIGURE 6.13 Luminance power vs. injection current for BGaN and GaN well LED structures.

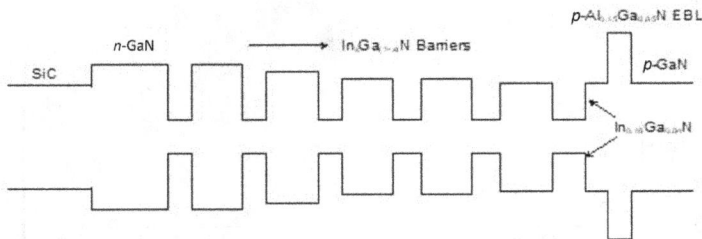

FIGURE 6.14 Outline structure of CSG-based MQW LED.

quantum efficiency, while comparing the structures with conventional structures. This enhancement is primarily due to the enhanced hole injection and modified band bend caused by the polarization effects. The outcomes describe an excellent agreement with the experimental data. Additionally, the lattice-matched SiC substrate technology increase the radiative recombination rate in the InGaN multiple quantum well light-emitting diode shown in Figs. 6.15 and 6.16.

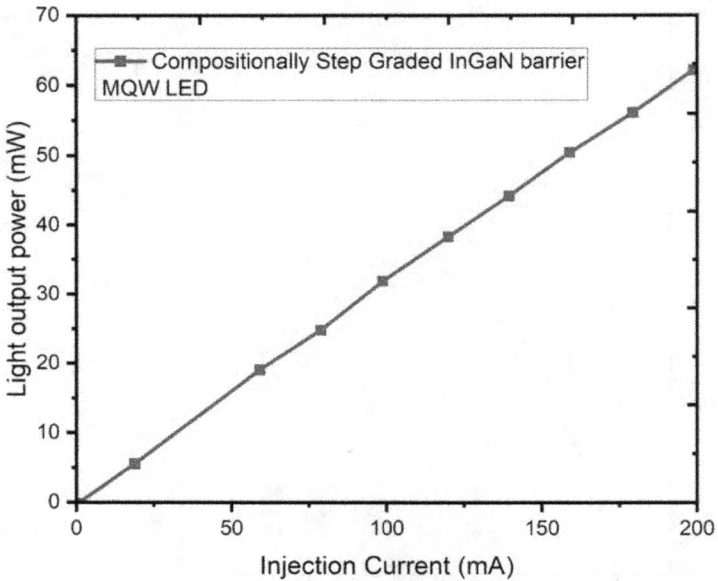

FIGURE 6.15   Injection current vs. light output power for CSG-based MQW LEDs.

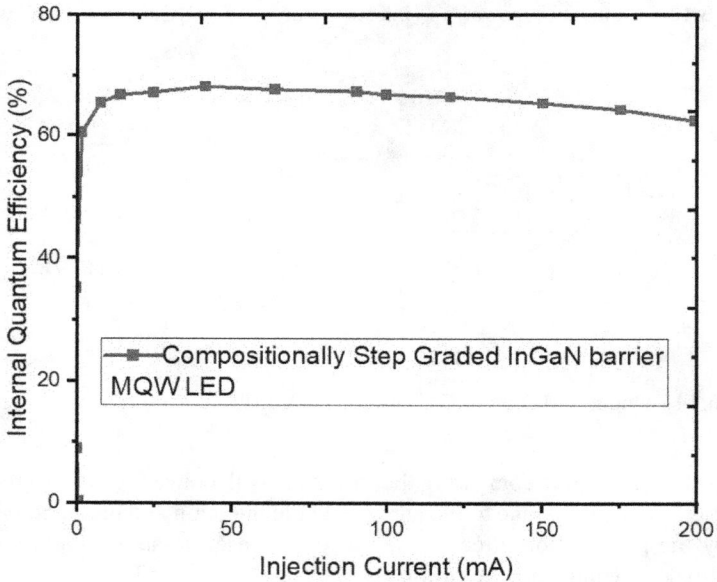

FIGURE 6.16   Injection current vs. internal quantum efficiency for CSG-based MQW LEDs.

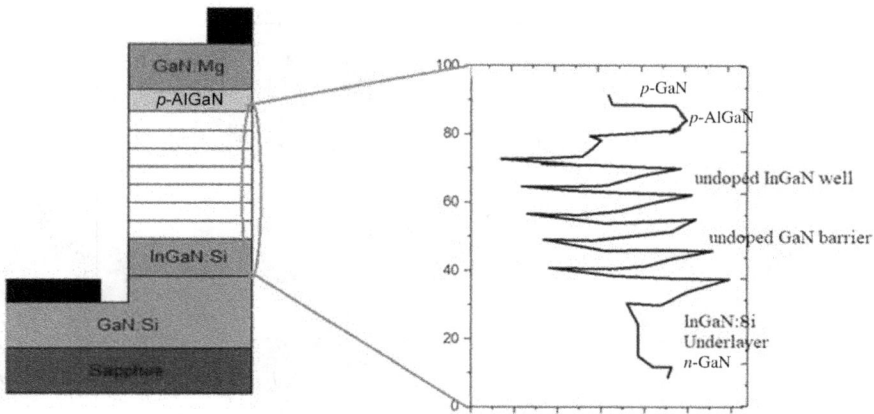

FIGURE 6.17 InGaN/GaN-based LED structures.

6.4.6 Efficiency Droop Reduction by Machine Learning Techniques

Rouet et al. [25] suggested a strategy that uses arithmetical and machine learning techniques to construct a light-emitting diode design process, by signifying new structures to sample in Figure 6.17 built in an extremely targeted way. He leverages Jones et al.'s [26] efficient global optimization (EGO) strategy, and iteratively chooses sample points to maximize the targeted efficiency enhancement while accounting for model uncertainty. The light-emitting diodes are provided with a "database" with a known structure, such as doping, the number of layers, widths, and composition labeled by ensuing electro luminescence IQE. The ML regression models are built to predict the efficiency of as yet unseen structures depicted in Figure 6.18. Subsequently, each new structure is sampled and added to the databases. Thus, the ML models are updated. In the approach to experimental design, one sample innovative LED structure for which the ML model predicts the greatest efficiencies is shown in Figure 6.19.

6.5 SUMMARY

This chapter has presented a comprehensive study of light extraction efficiency improvement techniques in a light-emitting diode. A range of characterization methodologies is introduced to unravel the different mechanisms for the recombination of carriers in GaN light-emitting diodes. To achieve the results, experimental setups that now can be used more widely to analyze LED wafer materials and devices had to be designed and built. The advancements for organic and inorganic blue LEDs are bright in the future due to the growth demand in visible communication, display, and solid-state lighting applications. The blue light-emitting diodes are considered the future utmost attractive solid-state lighting technology. Also, blue LEDs are considered a potential device for future generations of visible light communication applications.

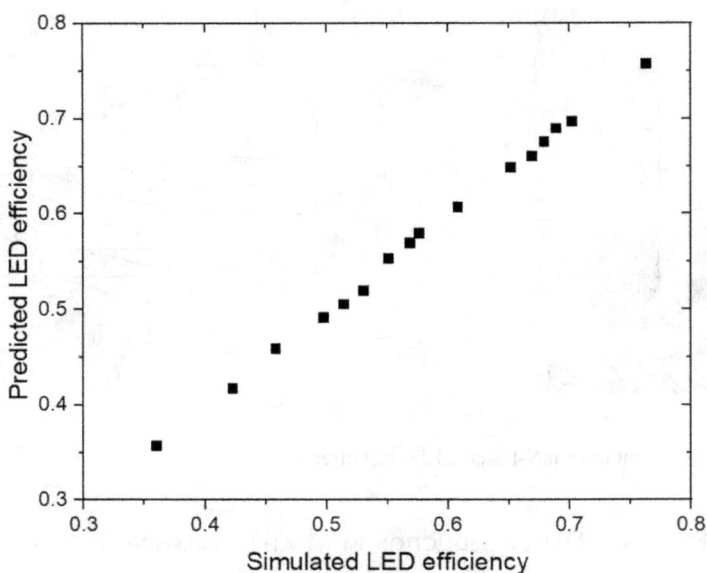

FIGURE 6.18 Machine learning algorithm predicted efficiencies vs. simulated efficiency for structures unseen by algorithms.

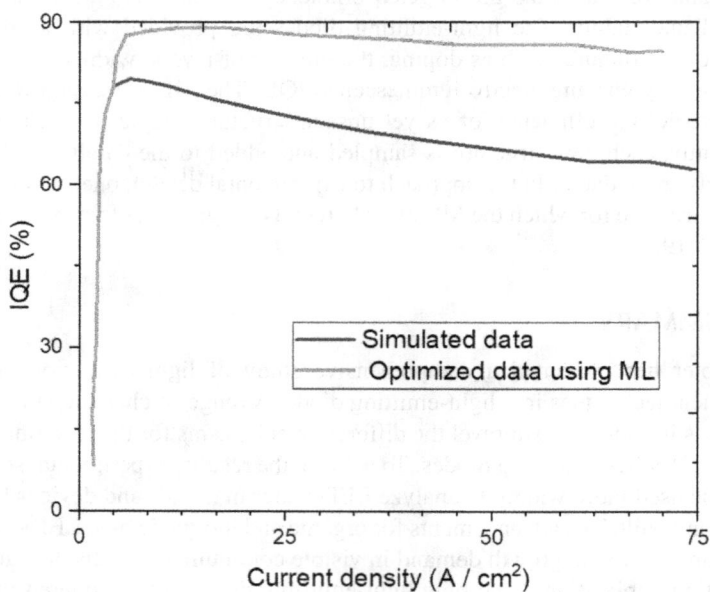

FIGURE 6.19 Comparison between simulation of an initial LED with machine learning optimized simulation of an LED.

REFERENCES

[1] Bergh A, Craford G, Duggal A, Haitz R. The promise and challenge of solid-state lighting. Physics Today. 2001 Dec 1; 54(12):42–7.

[2] Manikandan M, Nirmal D, Ajayan J, Mohankumar P, Prajoon P, Arivazhagan L. A review of blue light emitting diodes for future solid state lighting and visible light communication applications. Superlattices and Microstructures. 2019 Dec 1; 136:106294.

[3] Verma J, Islam SM, Verma A, Protasenko V, Jena D. 11.1 Light emitting diodes. Nitride semiconductor light-emitting diodes (LEDs): Materials, Technologies, and Applications. 2017 Oct 24:377.

[4] David A, Grundmann MJ. Droop in InGaN light-emitting diodes: A differential carrier lifetime analysis. Applied Physics Letters. 2010 Mar 8; 96(10):103504.

[5] Cho J, Schubert EF, Kim JK. Efficiency droop in light emitting diodes: Challenges and countermeasures. Laser & Photonics Reviews. 2013 May; 7(3):408–21.

[6] Kuo YK, Chang JY, Tsai MC. Enhancement in hole-injection efficiency of blue InGaN light-emitting diodes from reduced polarization by some specific designs for the electron blocking layer. Optics letters. 2010 Oct 1; 35(19):3285–7.

[7] Cheng LW, Xu CY, Sheng Y, Xia CS, Hu WD, Lu W. Study on GaN-based light emitting diode with InGaN/GaN/InGaN multi-layer barrier. Optical and Quantum Electronics. 2012 Jun; 44(3):75–81.

[8] Cheng L, Wu S. Performance enhancement of blue InGaN light-emitting diodes with a GaN–AlGaN–GaN last barrier and without an AlGaN electron blocking layer. IEEE Journal of Quantum Electronics. 2014 Feb 28; 50(4):261–6.

[9] Ngo TH, Gil B, Valvin P, Damilano B, Lekhal K, De Mierry P. Internal quantum efficiency in yellow-amber light emitting AlGaN-InGaN-GaN heterostructures. Applied Physics Letters. 2015 Sep 21; 107(12):122103.

[10] Jia C, Yu T, Lu H, Zhong C, Sun Y, Tong Y, Zhang G. Performance improvement of GaN-based LEDs with step stage InGaN/GaN strain relief layers in GaN-based blue LEDs. Optics Express. 2013 Apr 8; 21(7):8444–9.

[11] Lin RM, Li JC, Chou YL, Chen KH, Lin YH, Lu YC, Wu MC, Hung H, Lai WC. Improving the luminescence of InGaN–GaN blue LEDs through selective ring-region activation of the Mg-doped GaN layer. IEEE Photonics Technology Letters. 2007 May 29; 19(12):928–30.

[12] Cheng L, Wu S. Performance enhancement of blue InGaN light-emitting diodes with a GaN–AlGaN–GaN last barrier and without an AlGaN electron blocking layer. IEEE Journal of Quantum Electronics. 2014 Feb 28; 50(4):261–6.

[13] Saguatti D, Bidinelli L, Verzellesi G, Meneghini M, Meneghesso G, Zanoni E, Butendeich R, Hahn B. Investigation of efficiency-droop mechanisms in multi-quantum-well InGaN/GaN blue light-emitting diodes. IEEE Transactions on Electron Devices. 2012 Mar 15; 59(5):1402–9.

[14] Kuo YK, Tsai MC, Yen SH, Hsu TC, Shen YJ. Effect of p-type last barrier on efficiency droop of blue InGaN light-emitting diodes. IEEE Journal of Quantum Electronics. 2010 Apr 19; 46(8):1214–20.

[15] Yen SH, Tsai MC, Tsai ML, Shen YJ, Hsu TC, Kuo YK. Effect of n-type AlGaN layer on carrier transportation and efficiency droop of blue InGaN light-emitting diodes. IEEE Photonics Technology Letters. 2009 Jun 26; 21(14):975–7.

[16] Prajoon P, Nirmal D, Menokey MA, Pravin JC. A modified ABC model in InGaN MQW LED using compositionally step graded alternating barrier for efficiency improvement. Superlattices and Microstructures. 2016 Aug 1; 96:155–63.

[17] Prajoon P, Nirmal D, Menokey MA, Pravin JC. Efficiency enhancement of InGaN MQW LED using compositionally step graded InGaN barrier on SiC substrate. Journal of Display Technology. 2016 Oct 1; 12(10):1117–21.

[18] Prajoon P, Nirmal D, Menokey MA, Pravin JC. Temperature-dependent efficiency droop analysis of InGaN MQW light-emitting diode with modified ABC model. Journal of Computational Electronics. 2016 Dec; 15(4):1511–20.

[19] Verzellesi G, Saguatti D, Meneghini M, Bertazzi F, Goano M, Meneghesso G, Zanoni E. Efficiency droop in InGaN/GaN blue light-emitting diodes: Physical mechanisms and remedies. Journal of Applied Physics. 2013 Aug 21; 114(7):10_1.

[20] Hsu HC, Su YK, Huang SJ, Tseng CY, Cheng CY, Chen KC. Enhanced performance of nitride-based blue LED with step-stage MQW structure. IEEE Photonics Technology Letters. 2010 Dec 23; 23(5):287–9.

[21] Park, CH, Kang, SW, Jung, SG, Lee, DJ, Park, YW and Ju, BK, 2021. Enhanced light extraction efficiency and viewing angle characteristics of microcavity OLEDs by using a diffusion layer. Scientific Reports, 11(1):1–10.

[22] Yang X, Dev K, Wang J, Mutlugun E, Dang C, Zhao Y, Tan ST, Sun XW, Demir HV. Low-cost, large-scale, ordered ZnO nanopillar arrays for light extraction efficiency enhancement in quantum dot light-emitting diodes. In 2014 IEEE Photonics Conference 2014 Oct 12 (pp. 534–535). IEEE.

[23] Manikandan M, Nirmal D, Ajayan J, Arivazhagan L, Prajoon P, Dhivyasri G, Jagadeeswari M. Physics based modeling of AlGaN/BGaN quantum well based ultra violet light emitting diodes. Optical and Quantum Electronics. 2022 Mar; 54(3):1–3.

[24] Prajoon P, Nirmal D, Menokey MA, Pravin JC. Efficiency enhancement of InGaN MQW LED using compositionally step graded InGaN barrier on SiC substrate. Journal of Display Technology. 2016 Oct 1; 12(10):1117–21.

[25] Rouet-Leduc B, Barros K, Lookman T, Humphreys CJ. Optimisation of GaN LEDs and the reduction of efficiency droop using active machine learning. Scientific reports. 2016 Apr 26; 6(1):1–6.

[26] Jones DR, Schonlau M, Welch WJ. Efficient global optimization of expensive black-box functions. Journal of Global optimization. 1998 Dec; 13(4):455–92.

7 Efficiency Enhancement Techniques in Flexible and Organic Light-Emitting Diodes

J. Ajayan and T. D. Subash

CONTENTS

7.1   ORGANIC LIGHT-EMITTING DIODES (OLEDS): AN OVERVIEW

OLEDs are LEDs made of organic compounds having electro-luminescence (EL) properties that utilize electric current to produce light. EL is an electro-optic phenomenon in which light is emitted from a solid material or device through which current is passed. In the 1960s, it was discovered that organic materials can also produce EL. In 1977, Alan J. Heeger from University of California, USA, Alan Graham MacDiarmid from University of Pennsylvania, USA, and Hideki Shirakawa from University of Tsukuba, Japan, reported the development of conductive polymers that fueled the development of OLEDs. For the discovery and development of conductive polymers, they received the Nobel Prize in Chemistry in 2000. OLED display technology is widely used in applications like digital cameras, digital displays for smart phones, television screens, and automotive dash boards. OLEDs can be made thinner and lighter compared with liquid crystal displays (LCDs). In 1982, Ching Wan Tang and Steven Van Slyke invented OLED [1]. The OLED technology invented by Ching Wan Tang and Steven Van Slyke was first commercialized at the end of the 1990s. EL devices have been developed using a wide range of active thin film materials, such as sol-gel materials, molecular-doped polymers, conjugated polymers, and evaporated organic molecules. Polymer/organic material-based LEDs offer benefits, such as high brightness, low operating voltage, high speed response, wide viewing angle and

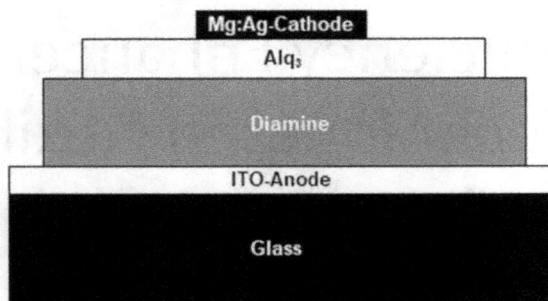

FIGURE 7.1   Organic EL cell.

Source: [4].

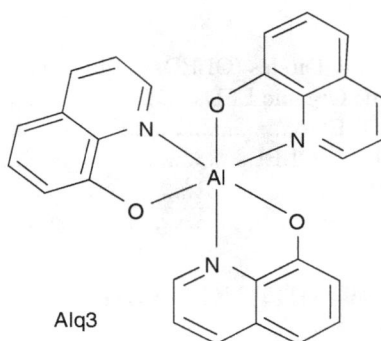

Alq3

FIGURE 7.2   The molecular structure of Alq$_3$ [4]. (Reprinted from [4] with permission from Elsevier.)

tenability to generate fundamental colors, inexpensive and easy fabrication process [2, 3]. The performance of OLEDs can be measured in Lumens/Watt.

In 1987, C. W. Tang and S. A. Van Slyke reported an organic EL diode with a brightness greater than 1000 Cd/m^2, luminous efficiency 1.5 lm/w, and external quantum efficiency (EQE) of 1%. The schematic of an organic EL cell is displayed in Figure 7.1. The organic EL cell consists of glass/ITO wafer, organic active layers (aromatic diamine and 8-hydroxyquinoline aluminum (Alq$_3$) and Mg:Ag top electrode. The molecular structure of Alq$_3$ is shown in Figure 7.2. The schematic of a multi-layer (ML) OLED is shown in Figure 7.3. The multi-layer OLED consists of an ITO/glass wafer, ITO anode, PEDOT:PSS hole transport enhancing layer, TPD (N, N–diphenyl-N, N–bis(3-methylphenyl)-1,1–biphenyl-4,4–diamine) hole transport layer (HTL), Alq$_3$ (aluminum tris(8-hydroxyquinoline)) electron transport layer (ETL) and LiF/Al cathode. A larger Alq$_3$ thickness is essential for enhancing the bandwidth of OLEDs. The reliability of an active matrix OLED array can be improved by introducing an amorphous-silicon (a-Si) thin film transistor (TFT) backplanes on flexible

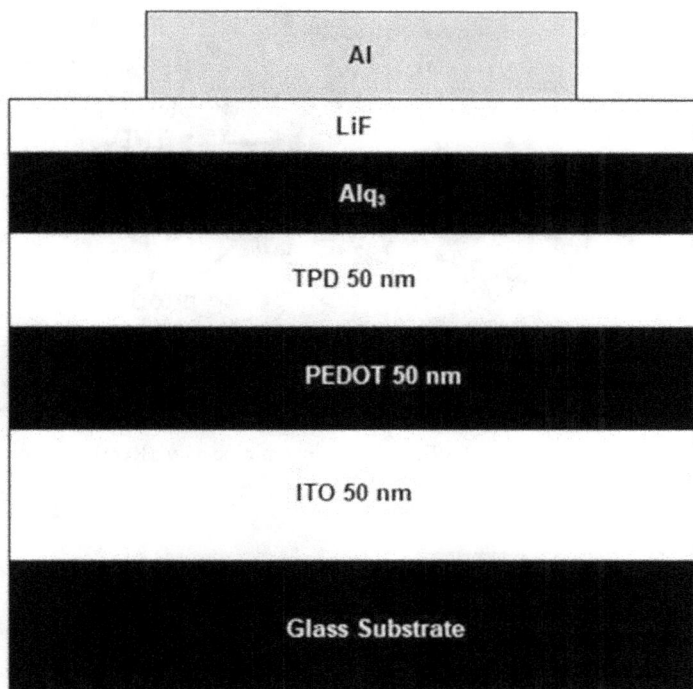

FIGURE 7.3   ML-OLED structure.

Source: [5].

and transparent plastic wafers [6, 7]. The introduction of TFTs helps to achieve low power consumption in OLEDs. However, stability of s-Si TFT is the most critical issue associated with the use of a-Si TFT backplanes in OLEDs on transparent and flexible wafers. The defect formation in a-Si and trapping of charges in the gate nitrides of a-Si TFTs leads to the increase of their threshold voltage with time [8–12].

Poor light extraction efficiency is one of the major disadvantages of OLEDs and therefore, it is essential to pay attention to the employment of optical microcavity structures in which the active layer must be located between the top and bottom mirrors to enhance the light extraction efficiency [13]. The utilization of microwave cavities along with distributed Bragg reflector (DBR) in OLEDs helps to improve the color tenability, enhancement of optical intensity, special redistribution of the emission, and spectral narrowing. The structure of an OLED with optical microcavity is shown in Figure 7.4. It consists of a ITO/glass wafer, Ag bottom metal mirror, organic HIL (hole injection layer), HTL, emitting layer, and an Al top metal mirror. The HILs are made of Alq_3, TPD, and CuPc (copper phthalocyanine).

It is required to develop high performance blue, green, and red OLEDs with long life time, good thermal and chemical stability and high EL efficiency [14, 15]. The schematic diagram of a microcavity blue OLED with TAT

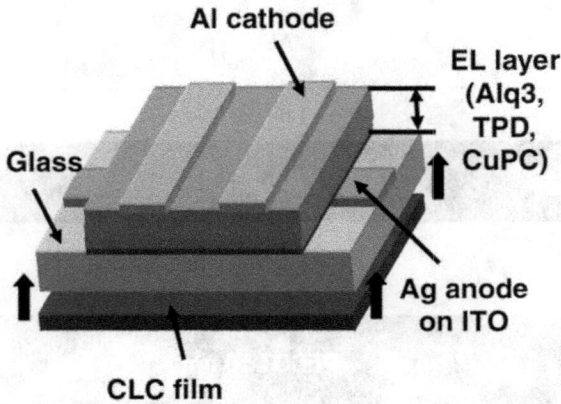

FIGURE 7.4 The structure of an OLED with optical microcavity. (Reprinted from Ref. [13] with permission from Elsevier.)

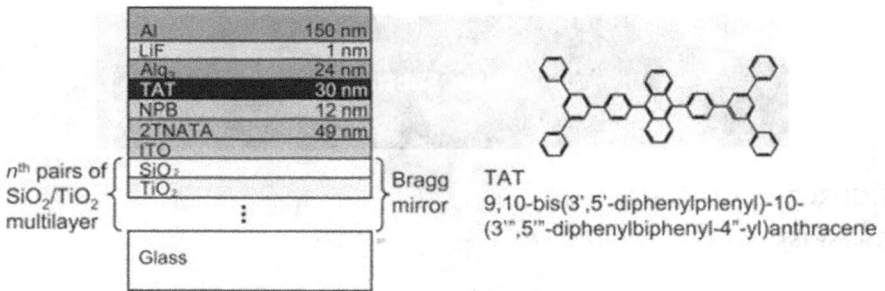

FIGURE 7.5 Microcavity blue OLED with TAT emitting material. (Reprinted from Ref. [16] with permission from Elsevier.)

(9,10-bis(3',5'–diphenylphenyl)-1'-(3''',5'''-diphenylbiphenyl-4''-yl)anthracene) emitting material is depicted in Figure 7.5. The key feature of this microcavity OLED is the SiO_2 (low-k)/TiO_2 (high-k) pair Bragg mirror. The electro-optical performance of TAT microcavity OLED is displayed in Figure 7.6. A TAT microcavity OLED with only one Bragg mirror, i.e., one pair SiO_2 (low-k)/TiO_2 (high-k) is most suitable for achieving high current density, luminescence, current efficiency, power efficiency, and EQE. OLEDs can be either the top emitting type or the bottom emitting type. The top emitting OLEDs are superior compared with the bottom emitting OLEDs because the top emitting OLEDs enable the fabrication of OLEDs on opaque and flexible wafers, such as metal foils, which are cheaper compared with flexible and transparent wafers, such as ITO [17]. PCR (pixel contrast ratio) and microcavity effects are the two critical issues for top emitting OLEDs. Silicon (Si) and silver (Ag) are the two most commonly used reflective anode materials for the development of top emitting OLEDs. Si exhibits low reflectivity, which commonly results in weak

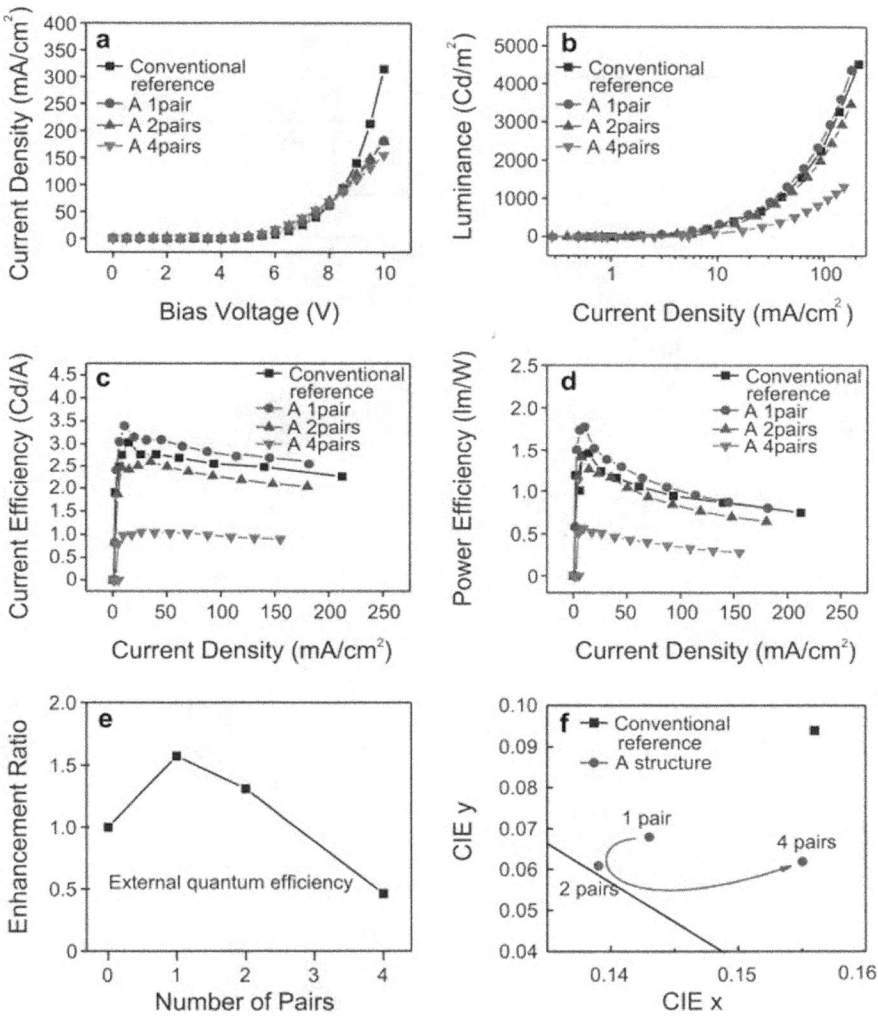

FIGURE 7.6 The electro-optical performance of TAT microcavity OLED. (Reprinted from Ref. [16] with permission from Elsevier.)

microcavity effects. A bottom electrode with high reflectivity is essential to achieve wide angle and multiple beam interferences within the cavity. On the other hand, bottom electrodes, such as Mo, Sm, and Cu with low reflectivity are considered as suitable for enhancing PCR.

The structures of Si-and Ag-based top emitting OLEDs are shown in Figure 7.7. Ag-based top emitting OLEDs exhibit superior luminescence compared with Si-based top emitting OLEDs (Figure 7.8). However, Si-based top emitting OLEDs outperform Ag-based top emitting OLEDs in terms of PCR.

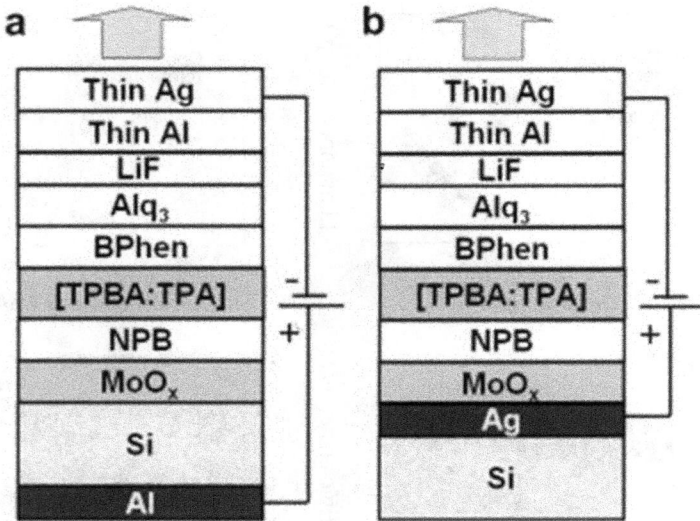

FIGURE 7.7 Structures of Si-and Ag-based top emitting OLEDs. (Reprinted from Ref. [18] with permission from Elsevier.)

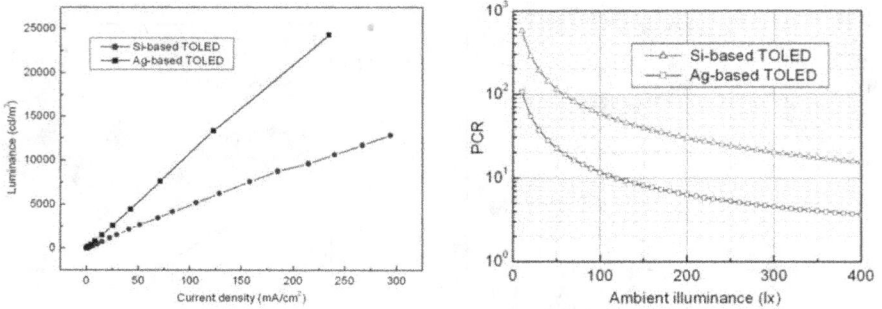

FIGURE 7.8 (a) Luminescence vs current density. (b) PCR vs ambient illuminance characteristics of Si-and Ag-based top emitting OLEDs. (Reprinted from Ref. [18] with permission from Elsevier.)

For the same current density, microcavity-based OLEDs provide superior brightness, EQE, and luminescence current efficiency compared with conventional OLEDs [19]. In 2014, An-Kai Ling et al. [20] reported a technique for enhancing the light out coupling efficiency (η_{out}) of OLEDs by blending a low refractive index polymer called PPFMA (poly(2,2,3,3,3-pentafluoropropyl methacrylate)) into the light-emitting layer (Figure 7.9). The influence of PPFMA concentration on the electro-optical characteristics of OLED structure is shown in Figure 7.10.

The OLED with 0.2% PPFMA exhibited superior current density, luminous efficiency, power efficiency, and EL intensity compared with OLEDs with 0.1%, 0.3%,

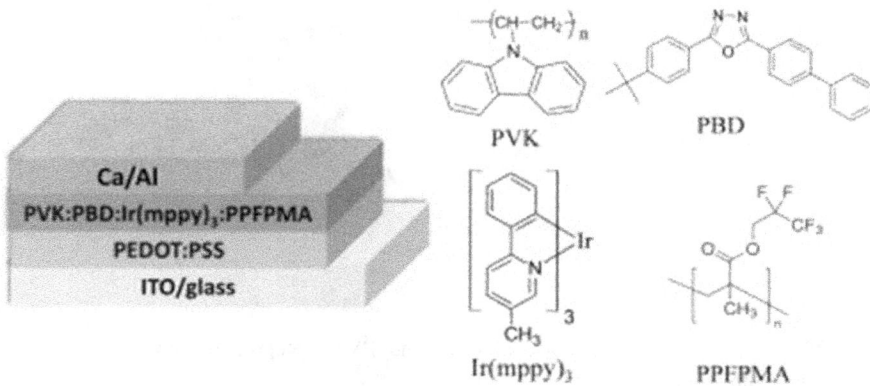

FIGURE 7.9 Light out-coupling efficiency enhancement of OLED using PPFMA. (Reprinted from Ref. [20] with permission from Elsevier.)

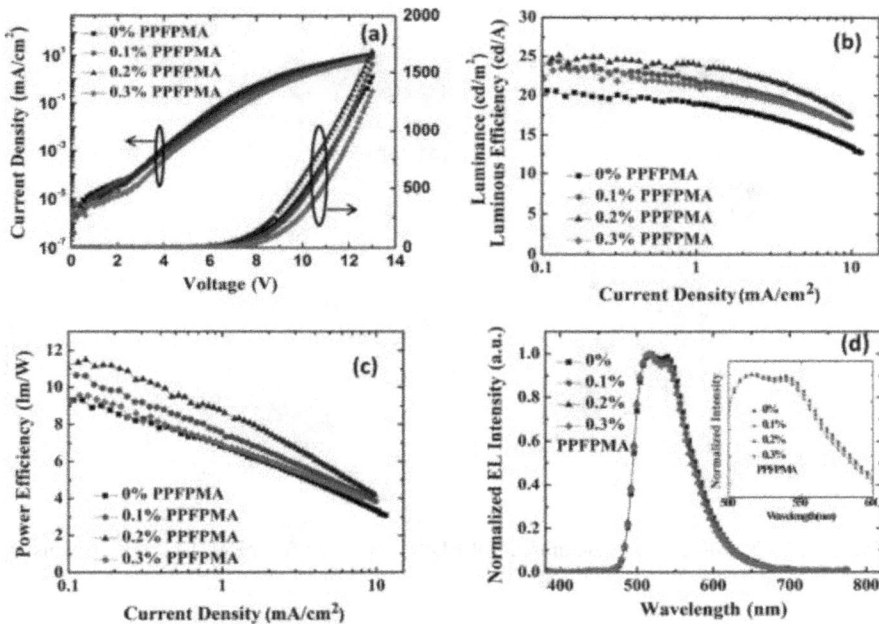

FIGURE 7.10 The influence of PPFMA concentration on the electro-optical characteristics of OLED using PPFMA. (Reprinted from Ref. [20] with permission from Elsevier.)

and 0% PPFMA concentration. An increase in PPFMA concentration leads to the increase of turn on voltage of the OLEDs. A 3:1 PVK (poly(N-vinyl carbazole)) and CBP (4,4'–bis(9-carbazolyl)-biphenyl) mixture can be used to create a broad band source profile to develop tunable microcavity OLEDs. The molecular structure of CBP and PVK is shown in Figure 7.11.

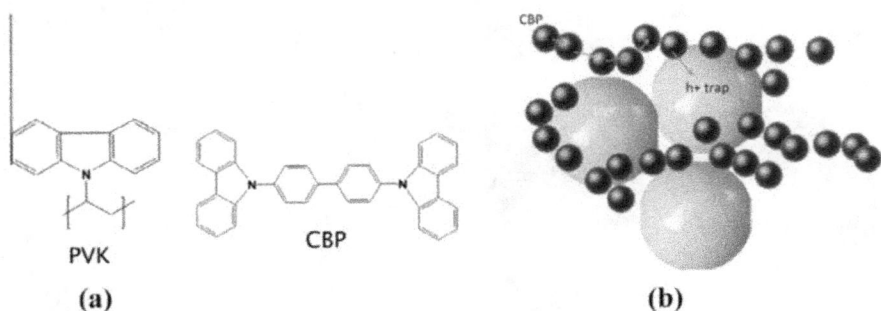

FIGURE 7.11 Molecular structure of CBP and PVK. (Reprinted from Ref. [21] with permission from Elsevier.)

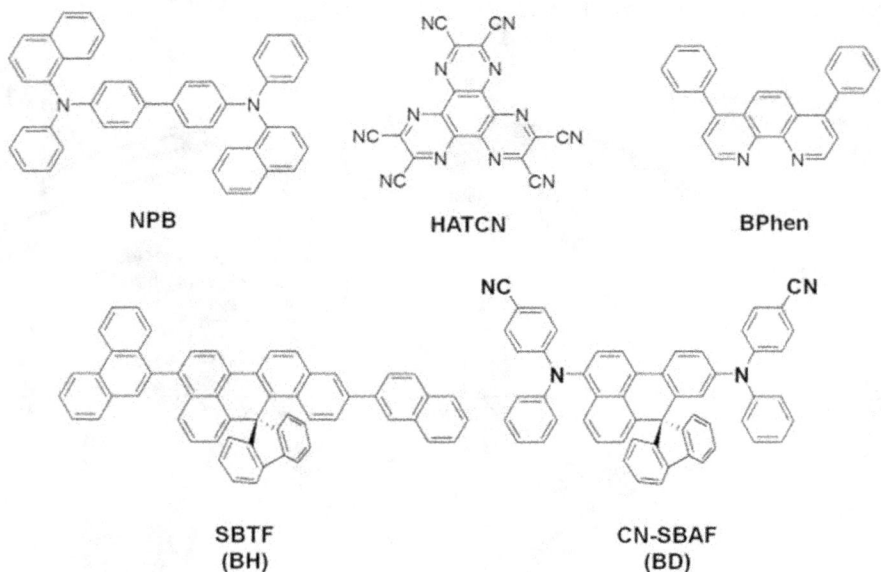

FIGURE 7.12 Molecular structures of OLEDs. (Reprinted from Ref. [22] with permission from Elsevier.)

In 2016, Min Chul Suh et al. [22] demonstrated that the viewing angle dependence in OLEDs can be effectively suppressed by adopting a nanoporous diffuser film on microcavity blue OLEDs. Figure 7.12 displays the organic materials used by Min Chul Suh et al. for developing the microcavity OLED. Min Chul Suh et al. compared the electro-optical performance of microcavity OLED with non-cavity OLED. The structure and electro-optical performance of OLEDs with and without cavity is depicted in Figure 7.13.

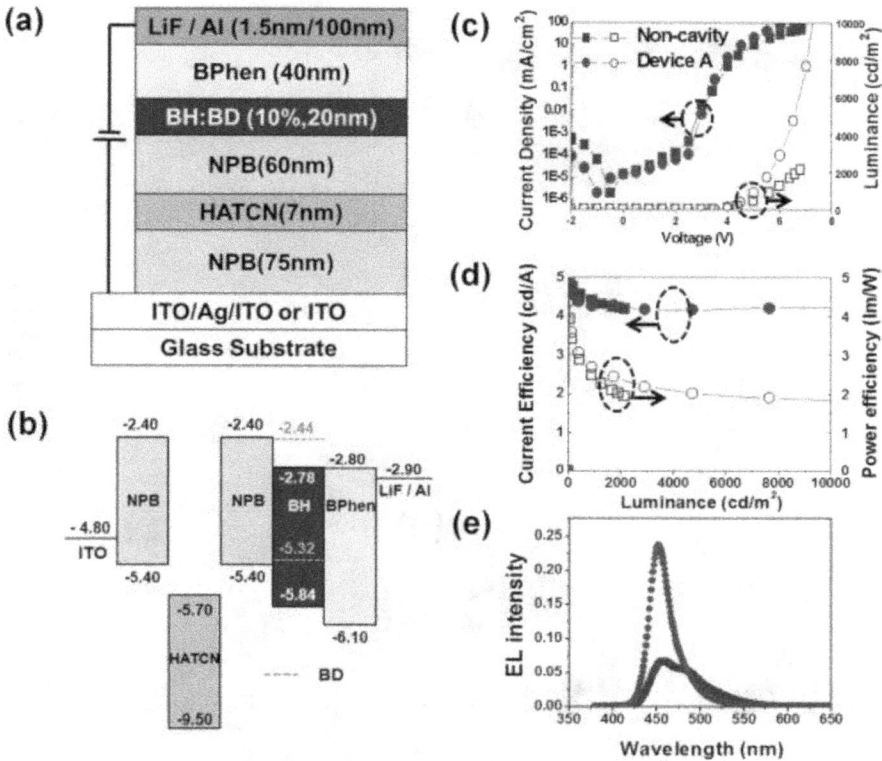

FIGURE 7.13 The structure and electro-optical performance of OLEDs with and without cavity OLEDs. (Reprinted from Ref. [22] with permission from Elsevier.)

A non-cavity OLED consist of 150 nm ITO, 75 nm NPB (N,N'-bis(naphthalen-1-yl)-N,N'-bis(phenyl)benzidine), 7 nm HATCN (1,4,5,8,9,11-hexaazatriphenylene-hexacarbonitrile), 1.5 nm LiF and 100 nm Al. Device A marked in Figure 7.13 consist of 1 nm glass/ITO, 17 nm Ag, 1 nm ITO, 75 nm NPB, 7 nm HATCN, 60 nm NPB, 20 nm 10% BH (SBTF) (10- (naphthalene-2-yl)-3-(phenanthrene-9-yl)spiro[benzo[ij] tetraphene-7,9'-fluorene]):BD   (CN-SBAF)   (N6,N9-bis(4-cyanophenyl)-N3,N9–diphenylspiro[benzo[de] anthracene-7,9'-fluorene]-3,9- diamine), 40 nm BPhen (4,7-diphenyl-1,10-phenanthroline), 1.5 nm LiF and 100 nm Al. The OLED with a microcavity exhibits relatively low turn on voltage, high current density, high luminescence, current and power efficiencies, and high EL intensity compared with non-cavity OLED. The structure and electro-optical performance of microcavity OLED with non-porous diffuser film is illustrated in Figure 7.14.

Device B indicated in Figure 7.14 consists of 1 nm diffuser/glass/ITO, 17 nm Ag, 1 nm ITO, 75 nm HATCN, 60 nm NPB, 20 nm 10% BH:BD, 40 nm BPhen, 1.5 nm LiF and 100 nm Al. Devices A and B exhibited the same turn on voltage but device A offers higher current and power efficiencies compared with device B. However,

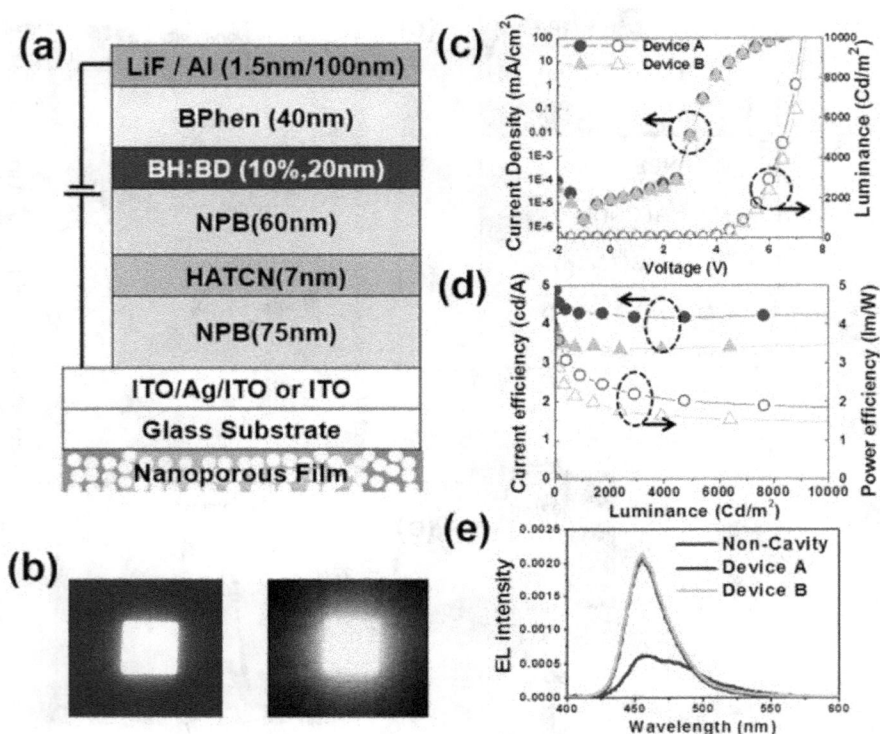

FIGURE 7.14 The structure and electro-optical performance of microcavity OLED with non-porous diffuser film. (Reprinted from Ref. [22] with permission from Elsevier.)

microcavity OLED with nano-porous diffuser film (device B) exhibited superior EL intensity. The structures of quantum dot (QD)/blue OLED with ITO and WO_3/ Ag/WO_3 (WAW) anodes are shown in Figure 7.15. The electro-optical performance of red QD/blue OLED is shown in Figure 7.16. In the plot 60, 80, 100, and 120 represents the thickness of ITO/W and WAW anode layers. Red QD/blue OLED with ITO anode exhibits superior luminescence, current efficiency, and EQE for a large ITO/W thickness of 120 nm. On the other hand, red QD/blue OLED with WAW anode exhibits superior luminescence and current efficiency for a WAW thickness of 100 nm compared with 60 nm, 80 nm, and 120 nm devices. But device exhibits superior EQE for a WAW thickness of 60 nm. In order to narrow and tune the emission spectrum of OLEDs, Fabry-Perot microcavity with a high Q-factor (quality factor) as shown in Figure 7.17 can be adopted.

Usually a nano-metal film is used as a semi-transparent optical output mirror electrode, which in comparison with a highly reflective back mirror electrode to form an optical microcavity. Pt, Au, Ag, and Al are widely used as semi-transparent anodes in microcavity OLEDs. Metal oxides such as MoO_3, WO_3, and V_2O_5 can be used as buffer materials to increase the anode metal work function in OLEDs. By adopting

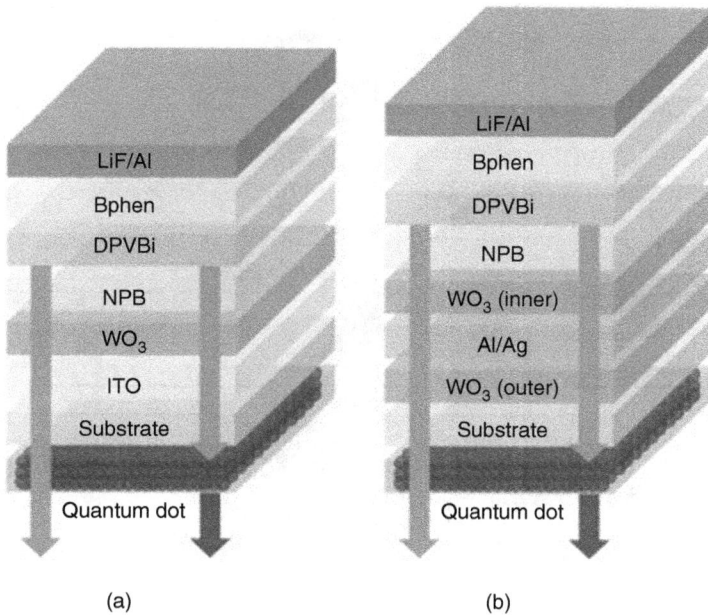

FIGURE 7.15   The structures of quantum dot (QD)/blue OLED with ITO and WO$_3$/Ag/WO$_3$ (WAW) anodes. (Reprinted from Ref. [23] with permission from Elsevier.)

metal nanoparticles, the luminescence of OLEDs can be effectively improved through surface plasmon (SP) effect. Surface plasmons are nothing but the oscillations of free electrons in nanometal particles at the dielectric/metal interfaces [25–32].

One method to enhance the color purity in OLEDs is to utilize the optical microcavity effects generated by the high reflectivity of electrodes. Microcavity structure usually consists of two metal mirrors and can be selectively extracted from the optical output with a specific wavelength equal to the length of the microcavity. DBR (distributed Bragg reflector) is made of alternately deposited high and low refractive index materials. SiO$_2$ (low-k)/TiO$_2$ (high-k) materials are usually used as DBR in OLEDs. The structure of conventional and inverted OLEDs is shown in Figure 7.18. In order to simplify the fabrication process and reduce the material cost hole transport/injection layer free solution processed OLEDs are highly preferred. Phosphoric acids react with the hydroxyl groups and bridges O$_2$ on the surface of ITO in the modification process as illustrated in Figure 7.19.

F$_5$BnPA (pentafluorobenzyl phosphonic acid) can be used as a modifier and EtOH (ethanol) and toluene can be used as solvents in solution processed OLEDs. The water contact angle obtained for ITO modified in toluene and EtOH is depicted in Figure 7.19(b). ITO modified in toluene obtained higher water contact angle compared with EtOH. Compared with ITO/Alq$_3$, ITO/NPB/Alq$_3$and F$_5$BnPA-ITO'/ Alq$_3$, F$_5$BnPA–ITO/Alq$_3$ structure offers high current density in solution processed

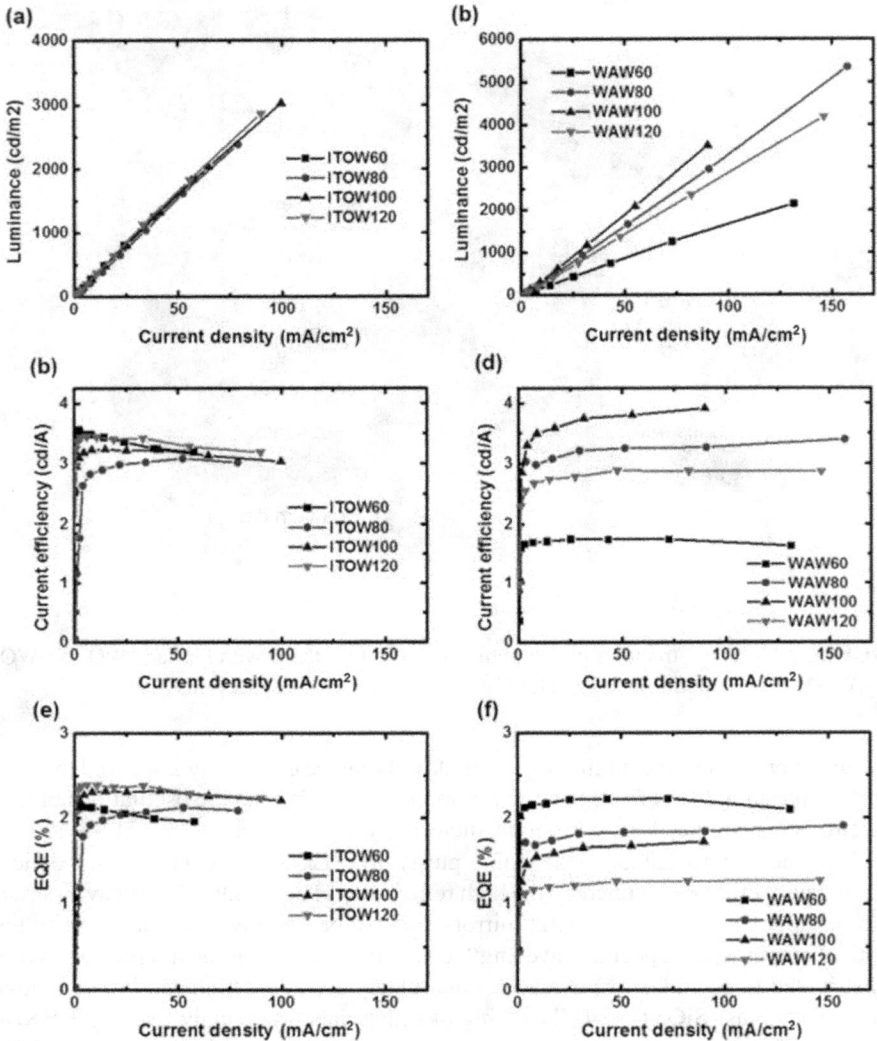

FIGURE 7.16 The electro-optical performance of red QD/blue OLEDs with ITO and WO_3/Ag/WO_3 (WAW) anodes. (Reprinted from Ref. [23] with permission from Elsevier.)

luminescence, current efficiency, and EL intensity. $HF_{21}DPA$, $HF_{17}DPA$, pCF_3BnPA, mpF_3BnPA, and F_5BnPA are the various phosphoric acids, which can be used as modifiers in solution processed OLEDs. The work function and contact angles of $HF_{21}DPA$, $HF_{17}DPA$, pCF_3BnPA, mpF_3BnPA, and F_5BnPA are -5.77 eV, -5.56 eV, -5.65 eV, -5.33 eV, and -5.20 eV, and 105.1^0, 103.2^0, 84.4^0, 82.2^0, 95.2^0, respectively. Among various phosphoric acid modified ITO OLEDs, F_5BnPA modified ITO

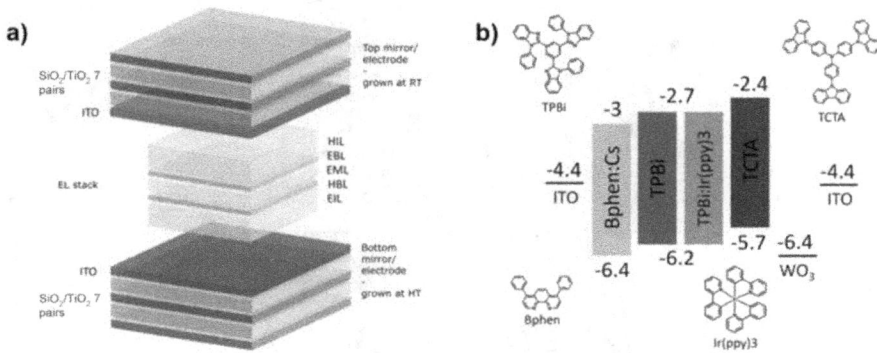

FIGURE 7.17 Fabry-Perot microcavity OLED with metal-free Bragg mirrors. (Reprinted from Ref. [24] with permission from Elsevier.)

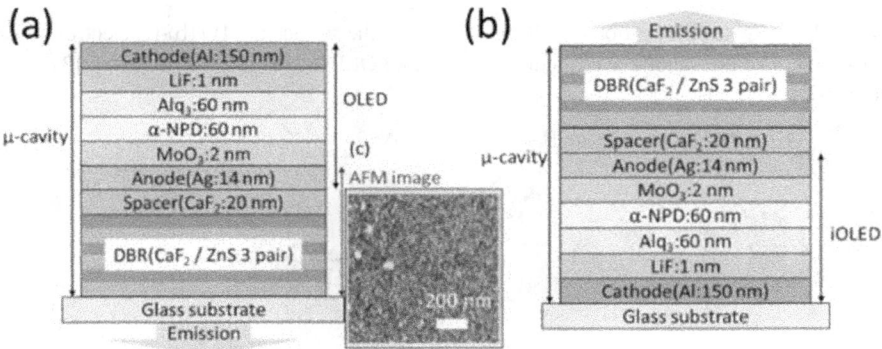

FIGURE 7.18 The structure of conventional and inverted OLEDs. (Reprinted from Ref. [33] with permission from Elsevier.)

device offers superior luminescence and current density. The schematic diagram and electro-optical performance of ITO/NPB/Rubrene and F_5BnPA/ITO/Rubrene OLEDs are illustrated in Figure 7.20. Compared with ITO/NPB/Rubrene OLED, F_5BnPA/ITO/Rubrene OLEDs provide superior luminescence, current efficiency, and power efficiency.

Consider the structure of an inverted UV OLED with ZnO EIL (electron injection layer) as shown in Figure 7.21. The effect of TAZ (3-(4-biphenyl)-4-phenyl-5-tert-butylphenyl-1,2,4-triazole) thickness variation on the electro-optical performance of UV OLED is depicted in Figure 7.22.

When TAZ layer thickness increases, EL intensity decreases, and EQE increases. The turn on voltage of UV OLED also increases with an increase in TAZ layer thickness.

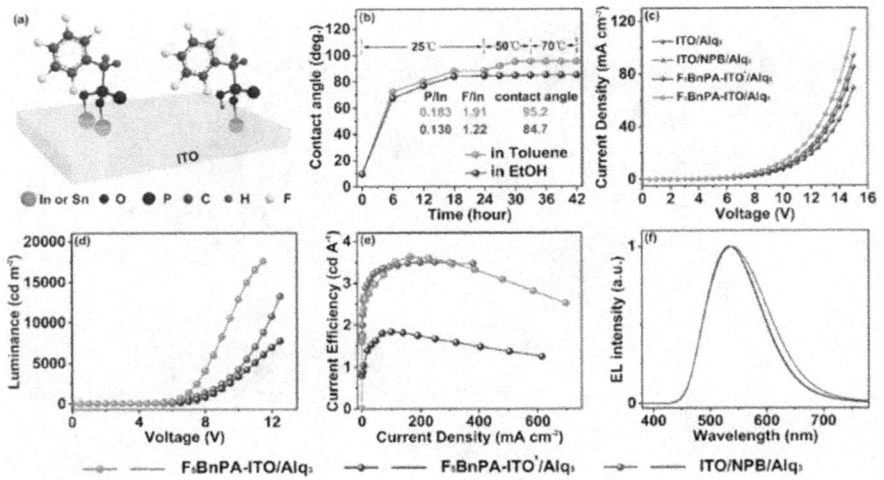

FIGURE 7.19 (a) Illustration of F_5BnPA binding to the surface of ITO (b)-(f) electro-optical performance of phosphoric acid modified ITO based OLEDs. (Reprinted from Ref. [34] with permission from Elsevier.)

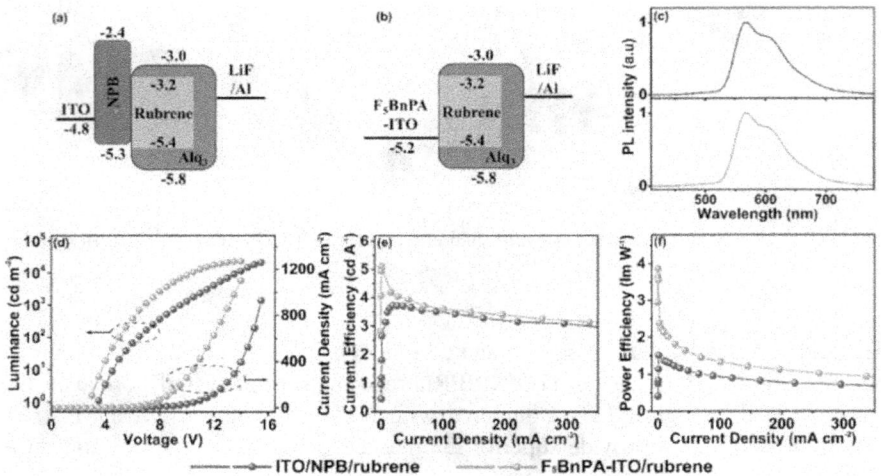

FIGURE 7.20 The schematic diagram and electro-optical performance of ITO/NPB/ rubrene and F_5BnPA/ITO/rubrene OLEDs. (Reprinted from Ref. [34] with permission from Elsevier.)

(a)

(b)

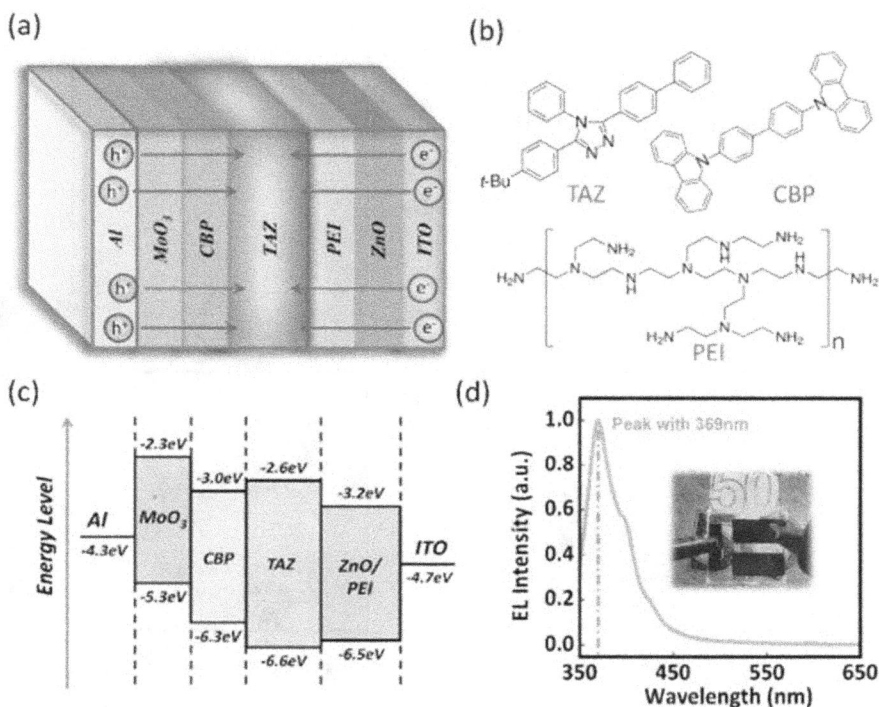

(c)

(d)

FIGURE 7.21 The structure of an inverted UV OLED with ZnO EIL. (Reprinted from Ref. [35] with permission from Elsevier.)

7.2 AN OVERVIEW OF BLUE ORGANIC LEDS

OLEDs are gaining attention due to their outstanding display image quality, market potentials in mobile, television, and laptop display applications. Therefore, commercialization activities are well on going. However, before OLED display technology can become the main stream technology living up to its complete potential and competing efficiently with LCD display technologies, several key stability and reliability related issues need to be resolved of which the development of blue OLEDs is the most critical one. It is essential to develop blue emitters for developing full-color OLED displays. di(styryl)arylene (DSA), diarylanthracene, pyrene, fluorine, 9, 10- (2-naphthyl) anthracene (ADN) with TBP (2, 5, 8, 11- (t–butyl)perylene) doping, TBADN (2-(t–butyl)-9,10- (2-naphthyl)anthracene), DTBADN (2,6- (t-butyl)-9,10- (2-naphthyl) anthracene), TTBADN (2,6- (t-butyl)-9,10- [6-(t-butyl)-2-naphthyl]anthracene), and MADN (2-methyl-9,10- (2-napthyl)anthracene) are considered as some of the stable blue emitting materials for the development of OLEDs [36]. TBP, IDE-102, and DSA-Ph are the widely used dopants for sky blue OLEDs. The power efficiency of fluorescent type blue OLEDs can be enhanced by doping two hole conduction layers of m-MTDATA (4, 4', 4''-Tris(N-3- methylphenyl-N-phenyl-amino) triphenylamine)

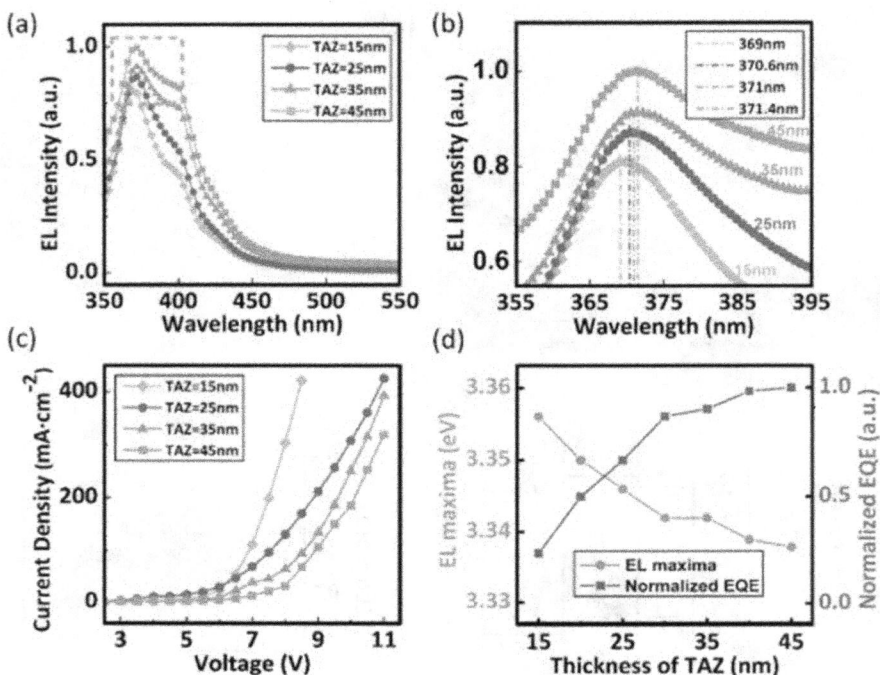

FIGURE 7.22 The effect of TAZ thickness variation on the electro-optical performance of UV OLED. (Reprinted from Ref. [35] with permission from Elsevier.)

and NPB (N, N'-diphenyl-N, N'-bis(1-naphthyl)-(1, 1'–biphenyl)-4, 4'-diamine) with F4-TCNQ (2,3,5,6-tetrafluoro-7,7,8,8-tetracyano-quinodimethane) as well as two ambipolar emitting layer with DSA Ph (p-bis(p-N, N-diphenylaminostyryl) benzene). The introduction of F4-TCNQ in the HTL/HIL layers efficiently minimizes the transport barrier at the NPB/ATA: F4-TCNQ/m-MTDATA [37]. The popular blue OLED architectures are shown in Figure 7.23. [36,38,39].

The first, second, and third generation blue OLEDs work based on fluorescence, phosphorescence, and thermally activated delayed fluorescence (TADF) mechanisms, respectively [36–43]. The EL efficiency in blue OLEDs can be effectively enhanced in three ways [40–42].

1. By using a hole blocking layer (HBL).
2. By using a composite hole transport layer (c-HTL).
3. By using a electron buffer layer (EBL) and hole buffer layer (HBL).

CuPc and MoO_3 are examples for EBL and HBL in blue OLEDs [38]. The current and power efficiencies in blue OLEDs can be adjusted by varying the thicknesses of EBL and HBL layers. Usually a thicker EBL and HBL is used to achieve higher current and power efficiencies.

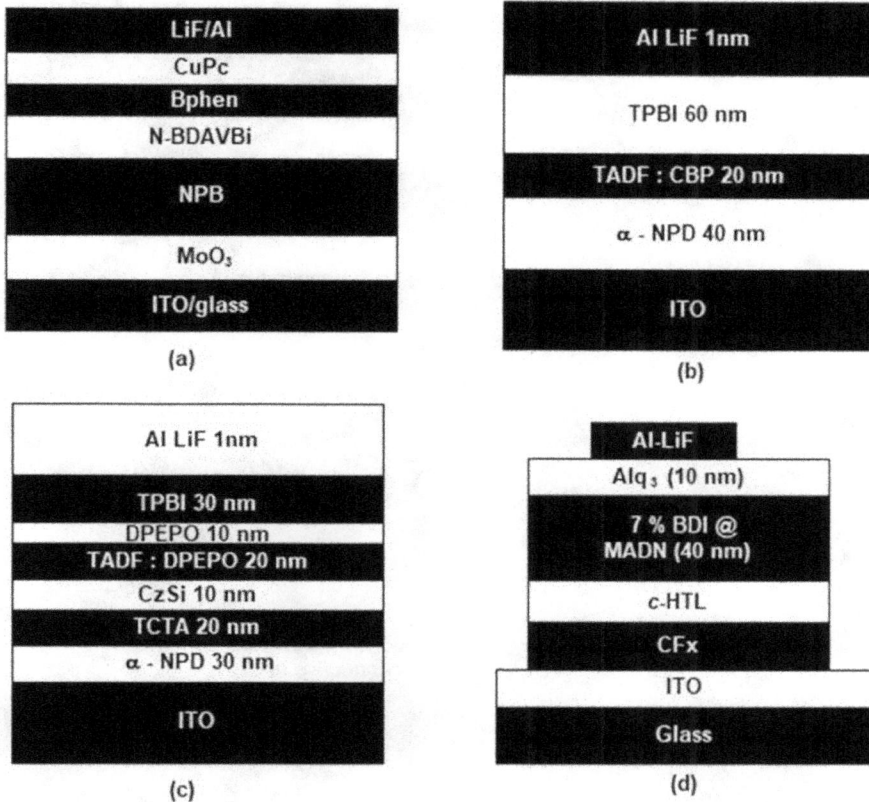

(a)

LiF/Al
CuPc
Bphen
N-BDAVBi
NPB
MoO₃
ITO/glass

(b)

Al LiF 1nm
TPBI 60 nm
TADF : CBP 20 nm
α - NPD 40 nm
ITO

(c)

Al LiF 1nm
TPBI 30 nm
DPEPO 10 nm
TADF : DPEPO 20 nm
CzSi 10 nm
TCTA 20 nm
α - NPD 30 nm
ITO

(d)

Al-LiF
Alq₃ (10 nm)
7 % BDI @ MADN (40 nm)
c-HTL
CFx
ITO
Glass

FIGURE 7.23 Popular blue OLED structures.

Source: [36–39].

7.3 INORGANIC FLEXIBLE OLEDS

Flexible inorganic LEDs (ILEDs) can be made in planar and non-planar layouts [44]. Flexible and stretchable ILEDs find applications in biomedicine including health monitors and oximeters. The rapid developments in high energy density flexible lithium ion battery propelled the development of flexible LEDs. The flexible batteries and flexible LEDs can be incorporated in future implantable or rollable electronics. One of the best flexible power sources are lithium ion batteries [45]. GaN-based flexible ILEDs are gaining tremendous interest due to their unique capabilities, such as high EQE, IQE, wide range of color emission, low energy consumption, and long life time. Usually, Si, SiC, and sapphire are used as the wafers for making GaN LEDs. But these wafers are rigid in nature. GaN-based ILEDs can be effectively grown on flexible wafers with the help of transfer printing technique and laser lift-off (LLO) process [46–49].

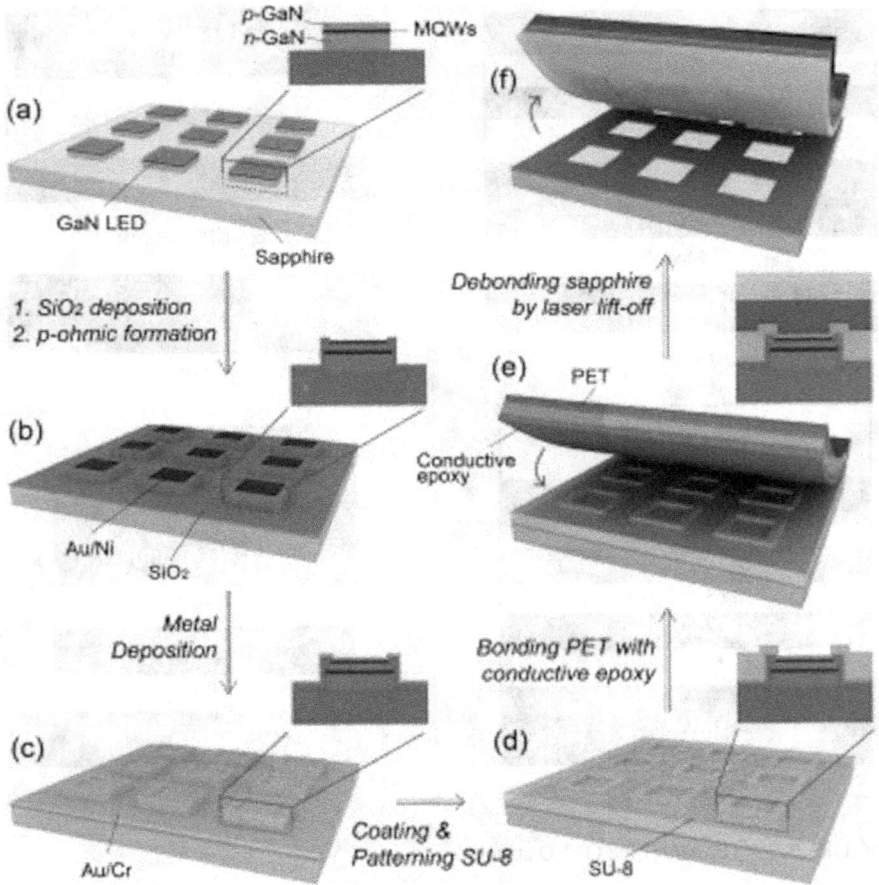

FIGURE 7.24 Demonstration of fabrication steps involved in transferring GaN ILED onto a PET film. (Reprinted from Ref. [55] with permission from Elsevier.)

The various steps involved in transferring GaN ILEDs grown on sapphire wafer is illustrated in [49]. The GaN LED on sapphire wafer consist of an intrinsic GaN layer of 2 µm thickness, 2 µm n-GaN layer, a few pairs of InGaN (3 nm)/GaN (7 nm) MQW (multiple quantum wells) and a 250 nm thick p-GaN. The fabrication process starts with opening the p-GaN and MQW layer by using the ICP-RIE technique to create n-contact regions. The p-ohmic contacts around the n-contact regions can be obtained by the use of electron beam evaporation technique in combination with LLO process. Then the sample can be annealed to enhance the ohmic contact properties. PECVD technique can be used to form a sacrificial layer of SiO$_2$ for the transfer printing process. In 2014, Won-Sik Choi et al. [50] reported the fabrication of a flexible InGaN ILED on 2-inch polyimide wafer using a direct transfer technique. However, the GaN on flexible wafers provide relatively low EL intensity and current density compared

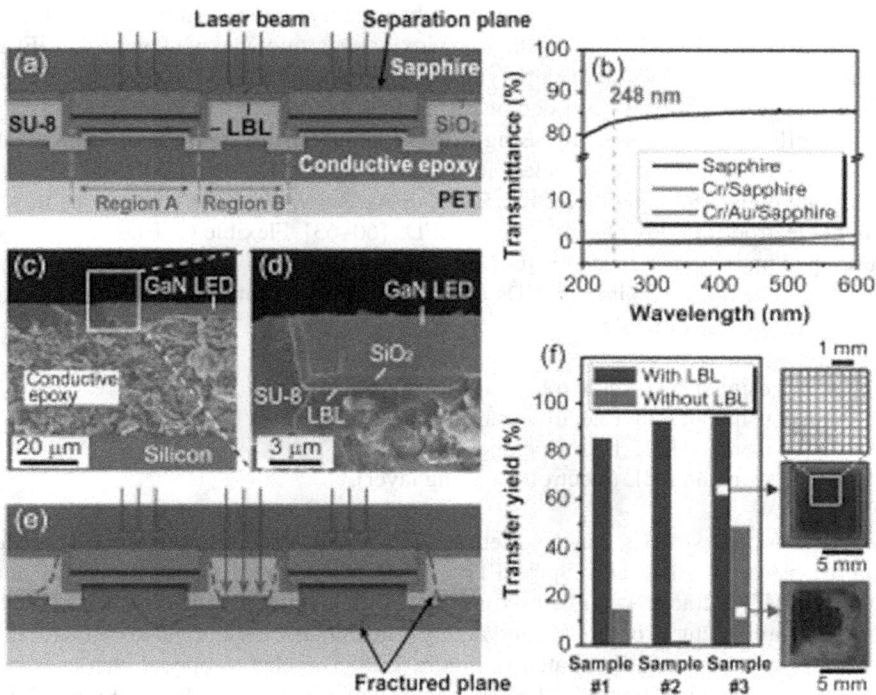

FIGURE 7.25 Demonstration of the LLO process to transfer GaN LED onto a PET wafer by adopting a Cr/Au LBL. (Reprinted from Ref. [55] with permission from Elsevier.)

with GaN on sapphire wafer [51–54]. The various fabrication steps involved in the transfer of GaN ILED on flexible PET wafer is illustrated in Figure 7.24 and Figure 7.25. Flexible ILEDs can also be made on ZnO nanowire (NW)/polyaniline (PANi) heterojunctions [56].

7.4   FLEXIBLE AND STRETCHABLE OLEDS

The main advantages of flexible OLED displays are (1) very thin; (2) unbreakable; (3) light weight; and (4) robust structures with arbitrary shapes. Various types of TFTs (thin film transistors) such as oxide semiconductor TFTs and Si-based TFTs have been demonstrated for the driving of active matrix (AM) flexible OLED displays. However, organic TFTs are most promising for driving AM flexible OLED displays due to their outstanding compatibility with flexible plastic wafers, capability to process large area flexible displays and flexible structures. In 2012, Yoshihide Fujisaki et al. [57] developed a flexible OLED-based AM OLED display using DNTT (dinaphtho[2, 3-b: 2', 3'-f]thieno[3, 2-b]- thiophene) organic semiconductor material. DNTT exhibits outstanding air stability and good hole mobility. In 2013, Matthew S. White et al. [58] developed a highly flexible and ultra thin stretchable polymer organic LED (POLED) that consists of PET (polyethylene terephthalate) substrate,

PEDOT:PSS (poly(3,4-ethylenedioxythiophene):poly (styrenesulphonate)) transparent electrode, AnE-PVstat (2-ethylhexyloxy side groups) active layer and LiF/Al metal electrode. AnE-PVstat is a polymer semiconductor with a bandgap of 2 eV. In 2016, Yi-Jun Wang et al. [59] demonstrated that the optical outcoupling efficiency can be efficiently enhanced by using a PIC (patterned inverted conical structure). ITO, DMD (dielectric/metal/dielectric Ex: ZnS/Ag/MoO$_3$), and DMDMD (dielectric/metal/dielectric/metal/dielectric, Ex: ZnS/Ag/ZnO/Ag/WO3) are used as transparent electrodes in flexible and organic LEDs [60–63]. Flexible OLEDs can also be developed on polymer substrates [64].

The efficiency of flexible OLEDs can be enhanced by introducing the following layers in the LED structure.

1. By using a HTL (hole transport layer).
2. By using an ETL (electron transport layer).
3. By using a HBL (hole blocking layer).
4. By using an EBL (electron blocking layer).

NBPhen doped with Cs can be used as ETL, BAlq can be used as HBL, NPB can be used as EBL and Spiro-TTB: F6-TCNNQ can be used as HTL and (NPB: Ir(MDQ)2(acac)) can be used as the emitting active layer in flexible OLEDs [65]. The architectures of flexible and highly efficient OLEDs fabricated on thermally stable mica substrate is illustrated in Figure 7.26. The electro-optical characteristic of highly efficient and flexible OLEDs on mica wafer is demonstrated in Figure 7.27. MGT device exhibited superior luminescence, current efficiency, and luminescence efficiency compared with G and MG OLEDs. However, power efficiency of MGT OLED is poor compared with G and MG OLEDs. The efficiency of flexible OLEDs can be enhanced by adopting a patterned Ag nanowire (Ag-NW) network transparent electrode. This enhancement in flexible OLED performance is due to the minimization of internal residual stress [67]. The outcoupling efficiency of flexible OLEDs can also be enhanced by using MLA (micro lens array) patterned parylene substrate and the various steps involved in the fabrication of flexible OLEDs on MLA patterned parylene wafer is shown in Figure 7.28.

STEP-1: Coating EGC-1700 (electronic grade coating-a fluorochemical acrylic polymer) on MLA film with a dip coater.
STEP-2: Parylene-C deposition on EGC-1700.
STEP-3: Formation of planarization layer.
STEP-4: Building OLED on MLA patterned parylene wafer.
STEP-5: Removal of glass wafer.

Flexible OLEDs on parylene wafer offers superior EL intensity, current efficiency, and current density compared with flexible OLEDs on glass wafer. The difference in refractive index between flexible wafer medium and air limits the performance of flexible OLEDs, which in turn minimizes the light out-coupling efficiency. This can be efficiently overcome by using an anti-reflective (AR) film in integration with

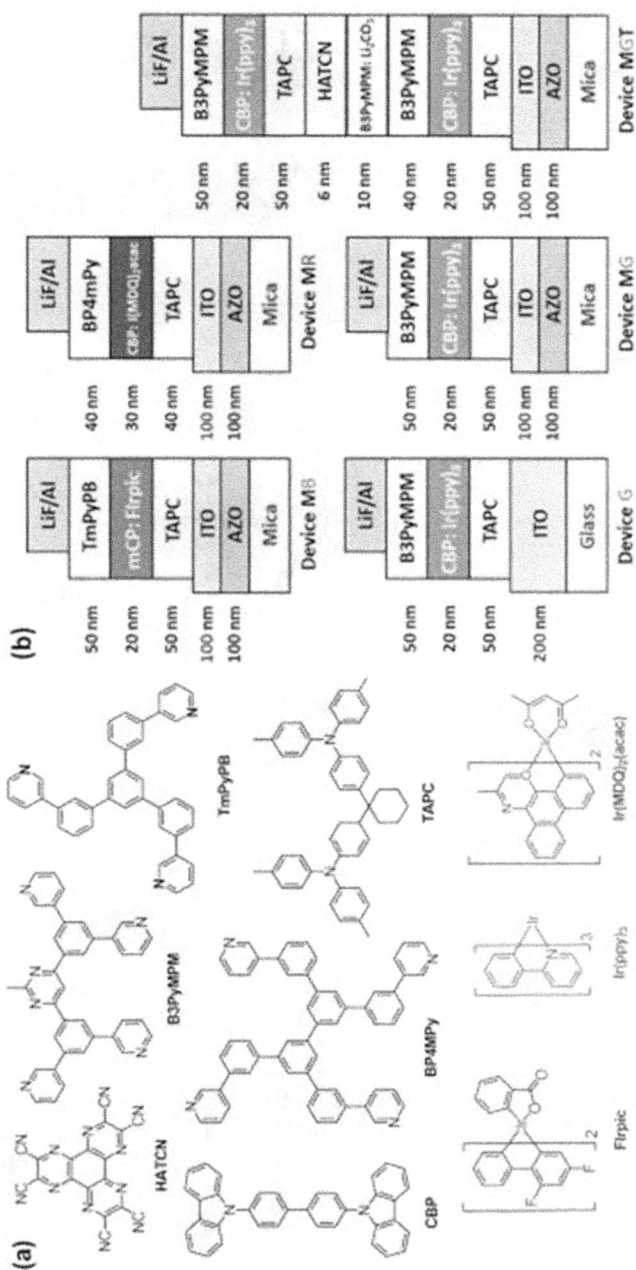

FIGURE 7.26 The architectures of flexible and highly efficient OLEDs fabricated on thermally stable mica substrate. (Reprinted from Ref. [66] with permission from Elsevier.)

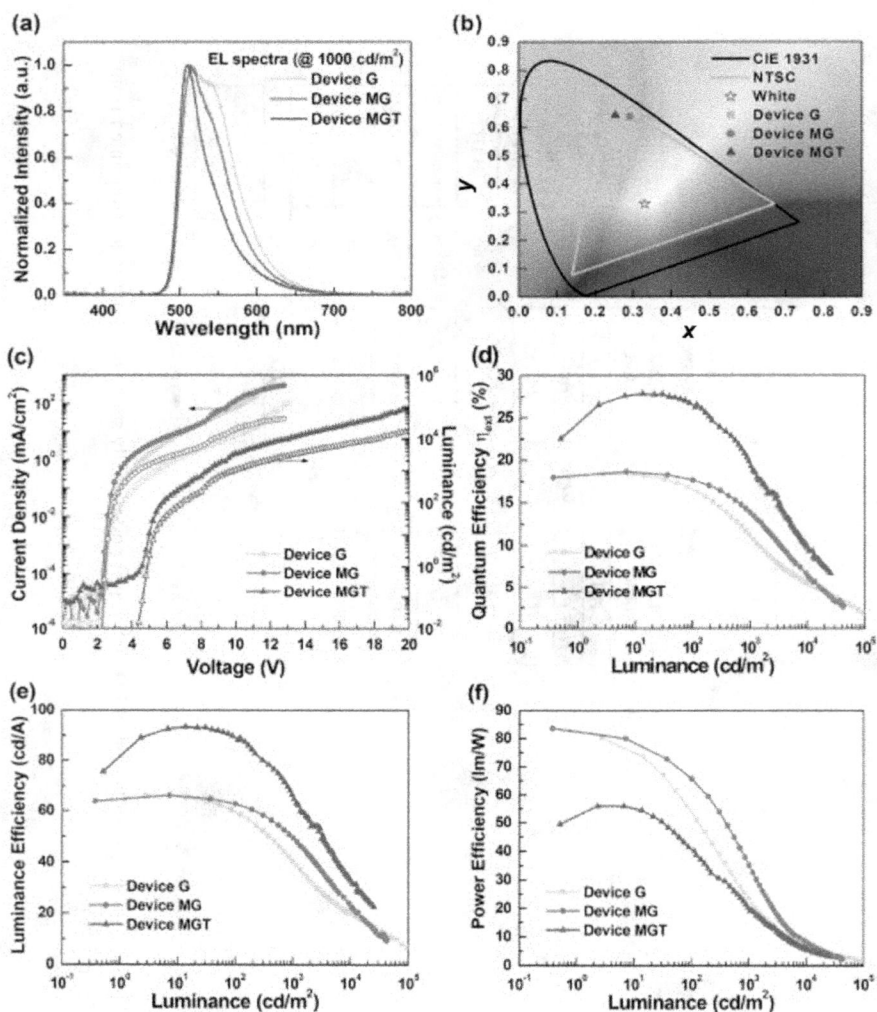

FIGURE 7.27 The electro-optical characteristics of highly efficient and flexible OLEDs on mica wafer. (Reprinted from Ref. [66] with permission from Elsevier.)

oxide/metal/oxide (OMO) structured electrode [69]. Flexible OLEDs with AR-OMO electrode offers superior luminous efficiency, EQE and EL intensity. The light extraction efficiency of flexible OLEDs can be significantly improved by introducing a scattering layer made of flexible polymer nano-particles [70]. The external light extraction efficiency can also be enhanced by introducing flexible porous films like PDMS (polydimethylsiloxane) [71]. Flexible OLEDs sometimes exhibit hysteresis effects due to the migration of carriers between the HTL layer and anode [72–74]. This can be eliminated by optimizing the O_2-plasma ratio at anode.

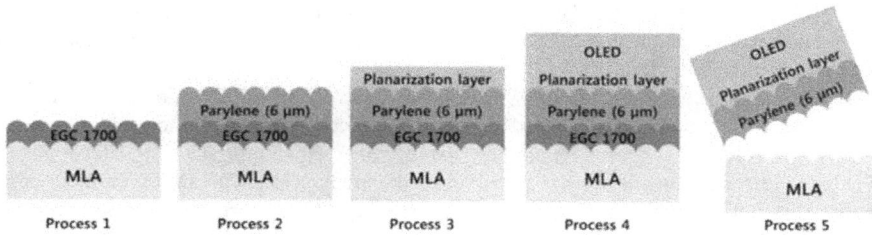

FIGURE 7.28   Various steps involved in the fabrication of flexible OLEDs on MLA patterned parylene wafer. (Reprinted from Ref. [68] with permission from Elsevier.)

7.5   SUMMARY

The unique characteristics of organic materials, especially their mechanically pliable and amorphous nature have enabled a wide variety of flexible optoelectronic devices, including solar cells, organic light-emitting diodes (OLEDs), organic sensors, neuromorphic devices, and electronic skin. The development of robust and mechanically flexible and durable OLEDs has inspired TVs and smartphones with curved displays, which are now entering the market. Apart from their use in displays, stretchable and flexible OLEDs enable a number of emerging new applications in which conformal integration or resilience against mechanical deformation are essential, e.g., for wearable and biomedical applications. In these applications, an ultrathin form factor is highly desirable to minimize volume and weight, enable ultimate conformability and mechanical flexibility importantly, reduce mechanical strain in the OLEDs upon folding and bending.

REFERENCES

[1]   C.W. Tang, Organic electroluminescent cell, US Patent No. 4356429,1982.
[2]   S.A. VanSlyke and C.W. Tang, Organic electroluminescent devices having improved power conversion efficiencies, US Patent No. 4539507A, 1985.
[3]   L. Ficke and M. Cahay, The bright future of organic LEDs, *IEEE Potentials*, Vol. 23, No. 5, pp. 31–34, Dec. 2003–Jan. 2004.
[4]   C.W. Tang and S.A. VanSlyke, Organic electroluminescent diodes, *Appl. Phys. Lett.,* Vol. 51, pp. 913–915, 1987.
[5]   A.Y. Mahmoud, K. Sulaiman, R.R. Bahabry, H. Alzahrani, Effect of Alq3 dopant on the optoelectronic parameters of NPD:Alq3 composite films, *Optical Materials*, Vol. 120, p. 111481, 2021.
[6]   H. Mu and D. Klotzkin, Measurement of electron mobility in Alq3 from optical modulation measurements in multilayer organic light-emitting diodes, *Journal of Display Technology*, Vol. 2, No. 4, pp. 341–346, Dec. 2006.
[7]   B. Hekmatshoar et al., Reliability of active-matrix organic light-emitting-diode arrays with amorphous silicon thin-film transistor backplanes on clear plastic, *IEEE Electron Device Letters*, Vol. 29, No. 1, pp. 63–66, Jan. 2008.
[8]   S. van Mensfoort, M. Carvelli, M. Megens et al., Measuring the light emission profile in organic light-emitting diodes with nanometre spatial resolution. *Nature Photon.,* Vol. 4, pp. 329–335 (2010).

[9] M. Wang, J. Lin, Y.C. Hsiao et al., Investigating underlying mechanism in spectral narrowing phenomenon induced by microcavity in organic light emitting diodes. *Nat Commun.*, Vol. 10, p. 1614 (2019).

[10] B. Krummacher, M.K. Mathai, V-.E. Choong, S.A. Choulis, F. So, Albrecht Winnacker, Influence of charge balance and microcavity effects on resultant efficiency of organic-light emitting devices, *Organic Electronics*, Vol. 7, No. 5, pp. 313–318, 2006.

[11] Hongmei Zhang, Jiawei Shi, Wei Wang, Shuxu Guo, Mingda Liu, HanYou, Dongge Ma, Tunability of resonant wavelength by Fabry–Perot microcavity in organic light-emitting diodes, *Journal of Luminescence*, Vol. 122–123, pp. 652–655, 2007.

[12] C.H. Cheung, A.M.C. Ng, A.B. Djurišić, Z.T. Liu, C.Y. Kwong, P.C. Chui, H.L. Tam, K.W. Cheah, W.K. Chan, J. Chan, A.W. Lu, A.D. Rakić, Angular dependence of the emission from low Q-factor organic microcavity light emitting diodes, *Displays*, Vol. 29, No. 4, pp. 358–364, 2008.

[13] Soon Moon Jeong, Yoichi Takanishi, Ken Ishikawa, Suzushi Nishimura, Goroh Suzaki, Hideo Takezoe, Sharply directed emission in microcavity organic light-emitting diodes with a cholesteric liquid crystal film, *Optics Communications*, Vol. 273, No. 1, pp. 167–172, 2007.

[14] Tyler Fleetham, Jeremy Ecton, Guijie Li, Jian Li, Improved out-coupling efficiency from a green microcavity OLED with a narrow band emission source, *Organic Electronics*, Vol. 37, pp. 141–147, 2016.

[15] Song Eun Lee, The Viet Hoang, Jeong-Hae Lee, Young Kwan Kim, Investigation of light out-coupling efficiency of blue OLED using microcavity effects, *Physica B: Condensed Matter*, Vol. 550, pp. 122–126, 2018.

[16] Hyoung Kun Kim, Sang-Hwan Cho, Jeong Rok Oh, Yong-Hee Lee, Jun-Ho Lee, Jae-Gab Lee, Soo-Kang Kim, Young-Il Park, Jong-Wook Park, Young Rag Do, Deep blue, efficient, moderate microcavity organic light-emitting diodes, *Organic Electronics*, Vol. 11, No. 1, pp. 137–145, 2010.

[17] Wenyu Ji, Letian Zhang, Zhang Tianyu, Wenfa Xie, Hanzhuang Zhang, High-contrast and high-efficiency microcavity top-emitting white organic light-emitting devices, *Organic Electronics*, Vol. 11, No. 2, pp. 202–206, 2010.

[18] Xiao-Wen Zhang, Hua-Ping Lin, Jun Li, Fan Zhou, Bin Wei, Xue-Yin Jiang, Zhi-Lin Zhang, Elucidations of weak microcavity effect and improved pixel contrast ratio in Si- based top-emitting organic light-emitting diode, *Current Applied Physics*, Vol.12, No. 5, pp. 1297–1301, 2012.

[19] Fengying Ma, Jianpo Su, Maotian Guo, Qiaoxia Gong, Zhiyong Duan, Jing Yang, Yanli Du, Bin Yuan, Xingyuan Liu, Model and simulation on the efficiencies of microcavity OLEDs, *Optics Communications*, Vol. 285, No. 13–14, pp. 3100–3103, 2012.

[20] An-Kai Ling, Chun-Hao Lin, Hsun Liang, Fang-Chung Chen, Tunable microcavities in organic light-emitting diodes by way of low-refractive-index polymer doping, *Organic Electronics*, Vol. 15, No. 12, pp. 3648–3653, 2014.

[21] Emily S. Hellerich, Eeshita Manna, Robert Heise, Rana Biswas, Ruth Shinar, Joseph Shinar, Deep blue/ultraviolet microcavity OLEDs based on solution-processed PVK:CBP blends, *Organic Electronics*, Vol. 24, pp. 246–253, 2015.

[22] Min Chul Suh, Beom Pyo, Hyung Suk Kim, Suppression of the viewing angle dependence by introduction of nanoporous diffuser film on blue OLEDs with strong microcavity effect, *Organic Electronics*, Vol. 28, pp. 31–38, 2016.

[23] Jung HyukIm, Kyung-TaeKang, Jong Sun Choi, Kwan Hyun Cho, Strong microcavity effects in hybrid quantum dot/blue organic light-emitting diodes using Ag based electrode, *Journal of Luminescence*, Vol. 203, pp. 540–545, 2018.

[24] Armando Genco, Goffredo Giordano, Sonia Carallo, Gianluca Accorsi, Yu Duan, Salvatore Gambino, Marco Mazzeo, High quality factor microcavity OLED employing metal-free electrically active Bragg mirrors, *Organic Electronics*, Vol. 62, pp. 174–180, 2018.

[25] Yadong Liu, Yan Zhao, Haihao Zhang, Xinchao Song, Jihua Zhou, Zhisheng Wu, Jie Zhang, Wang jun Guo, Yuhua Mi, The application of the nanostructure aluminum in the blue organic light-emitting devices, *Organic Electronics*, Vol. 57, pp. 1–6, 2018.

[26] Minqiang Wan, Wenqing Zhu, Lu Huang, Yunping Zhao, Zixing Wang, Jun Li, Bin Wei, Improving color rendering index of top-emitting white OLEDs with single emitter by using microcavity effects, *Organic Electronics*, Vol. 100, p. 106381, 2022.

[27] Dong Jun Lee, Soo Jong Park, Cheol Hwee Park, Young Wook Park, Byeong-Kwon Ju, Microcavity characteristics analysis of micro-shuttered organic light-emitting diodes, *Thin Solid Films*, Vol. 692, p. 137643, 2019.

[28] Takuya Kitabayashi, Teruyuki Asashita, Naoya Satoh, Takayuki Kiba, Midori Kawamura, Yoshio Abe, Kyung Ho Kim, Fabrication and characterization of microcavity organic light-emitting diode with CaF2/ZnS distributed Bragg reflector, *Thin Solid Films*, Vol. 699, p. 137912, 2020.

[29] Chih-Wei Huang, Ting-An Lin, Wei-Kai Lee, Chen-Han Lu, Tanmay Chatterjee, Chin-Hui Chou, Ken-Tsung Wong, Chung-Chih Wu, Analyses of emission efficiencies of white organic light-emitting diodes having multiple emitters in single emitting layer, *Organic Electronics*, Vol. 104, p. 106474, 2022.

[30] Pasquale Cusumano, Giovanni Garraffa, Salvatore Stivala, A simple method for the photometric characterization of organic light-emitting diodes, *Solid-State Electronics*, Vol. 195, p. 108394, 2022.

[31] Amani Ouirimi, Alex Chamberlain Chime, Nixson Loganathan, Mahmoud Chakaroun, Alexis P.A. Fischer, Daan Lenstra, Threshold estimation of an organic laser diode using a rate-equation model validated experimentally with a microcavity OLED submitted to nanosecond electrical pulses, *Organic Electronics*, Vol. 97, p. 106190, 2021.

[32] Hyunsu Cho, Chul Woong Joo, Sukyung Choi, Chan-mo Kang, Byoung-Hwa Kwon, Jin- Wook Shin, Kukjoo Kim, Dae-Hyun Ahn, Nam Sung Cho, Gi Heon Kim, Highly conductive and transparent thin metal layer for reducing microcavity effect in top-emitting white organic light-emitting diode, *Organic Electronics*, Vol. 106, p. 106537, 2022.

[33] Takuya Kitabayashi, Teruyuki Asashita, Naoya Satoh, Takayuki Kiba, Midori Kawamura, Yoshio Abe, Kyung Ho Kim, Fabrication and characterization of microcavity organic light-emitting diode with CaF2/ZnS distributed Bragg reflector, *Thin Solid Films*, Vol. 699, p. 37912, 2020.

[34] Fei Huang, Hongli Liu, Xianggao Li, Shirong Wang, Highly efficient hole injection/transport layer-free OLEDs based on self-assembled monolayer modified ITO by solution-process, *Nano Energy*, Vol. 78, p. 105399, 2020.

[35] Jin Na, Sheng Bi, Chengming Jiang, Jinhui Song, Achieving the hypsochromic electroluminescence of ultraviolet OLED by tuning excitons relaxation, *Organic Electronics*, Vol. 82, p. 105718, 2020.

[36] Shih-Wen Wen, Meng-Ting Lee and C.H. Chen, Recent development of blue fluorescent OLED materials and devices, *Journal of Display Technology*, Vol. 1, No. 1, pp. 90–99, Sept. 2005.

[37] H.M. Zhang, W.C.H. Choy and K. Li, Blue organic LEDs with improved power efficiency, *IEEE Transactions on Electron Devices*, Vol. 57, No. 1, pp. 125–128, Jan. 2010.

[38] Z.Q. Jiao et al., Enhancing performances of blue organic light-emitting devices incorporating CuPc as electron buffer layers, *IEEE Electron Device Letters*, Vol. 33, No. 3, pp. 408–410, March 2012.

[39] Zhang, Q., Li, B., Huang, S. et al., Efficient blue organic light-emitting diodes employing thermally activated delayed fluorescence. *Nature Photon.,* Vol. 8, pp. 326–332, 2014.

[40] Kim, M., Jeon, S.K., Hwang, S.-H. and Lee, J.Y. Stable blue thermally activated delayed fluorescent organic light-emitting diodes with three times longer lifetime than phosphorescent organic light-emitting diodes. *Adv. Mater.*, Vol. 27, pp. 2515–2520, 2015.

[41] Miwa, T., Kubo, S., Shizu, K. et al., Blue organic light-emitting diodes realizing external quantum efficiency over 25% using thermally activated delayed fluorescence emitters. *Sci Rep.,* Vol. 7, p. 284, 2017.

[42] Lee, S.Y., Adachi, C. and Yasuda, T. High-efficiency blue organic light-emitting diodes based on thermally activated delayed fluorescence from phenoxaphosphine and phenoxathiin derivatives. Adv. Mater., Vol. 28, pp. 4626–4631, 2016.

[43] M. Manikandan, D. Nirmal, J. Ajayan, P. Mohankumar, P. Prajoon, L. Arivazhagan, A review of blue light emitting diodes for future solid state lighting and visible light communication applications, *Superlattices and Microstructures*, Vol. 136, p. 106294, 2019.

[44] T. Kim, R. Kim and J.A. Rogers, Microscale inorganic light-emitting diodes on flexible and stretchable substrates, *IEEE Photonics Journal*, Vol. 4, No. 2, pp. 607–612, April 2012.

[45] Anonymous, Bendable battery yields flexible LED. *Nature,* Vol. 490, p. 9, 2012.

[46] M.K. Kelly, O. Ambacher, R. Dimitrov, R. Handschuh, and M. Stutzmann, Optical process for liftoff of group III-nitride films, *Phys. Status Solidi A.*, Vol. 159, No. 1, pp. R3–R4, 1997.

[47] W.S.Wong, T. Sands, and N.W. Cheung, Damage-free separation of GaN thin films from sapphire substrates, *Appl. Phys. Lett.*, Vol. 72, No. 5, pp. 599–601, 1998.

[48] W.S. Wong, M. Kneissl, P. Mei, D.W. Treat, M. Teepe, and N.M. Johnson, Continuous-wave InGaN multiple-quantum-well laser diodes on copper substrates, *Appl. Phys. Lett.*, Vol. 78, No. 9, pp. 1198–1200, 2001.

[49] J. Chun et al., Transfer of GaN LEDs from sapphire to flexible substrates by laser lift-off and contact printing, *IEEE Photonics Technology Letters*, Vol. 24, No. 23, pp. 2115–2118, Dec. 1, 2012.

[50] W. Choi, H.J. Park, S. Park and T. Jeong, Flexible InGaN LEDs on a polyimide substrate fabricated using a simple direct-transfer method, *IEEE Photonics Technology Letters*, Vol. 26, No. 21, pp. 2115–2117, 1 Nov.1, 2014.

[51] Lee, H.E., Choi, J., Lee, S.H., Jeong, M., Shin, J.H., Joe, D.J., Kim, D., Kim, C.W., Park, J.H., Lee, J.H., Kim, D., Shin, C.S., Lee, K.J., *Adv. Mater.* Vol. 30, p. 1800649, 2018.

[52] Tian Z., Li Y., Su X., Feng L., Wang S., Ding W., Li Q., Zhang Y., Guo M., Yun F., Lee S.W.R. Super flexible GaN light emitting diodes using microscale pyramid arrays through laser lift-off and dual transfer. Opt Express. 2018 Jan 22, Vol. 26, No. 2, pp. 1817–1824, 2018.

[53] G. Tabares et al., Impact of the bending on the electroluminescence of flexible InGaN/GaNlight-emitting diodes, *IEEE Photonics Technology Letters*, Vol. 28, no. 15, pp. 1661–1664, 1 Aug. 1, 2016.

[54] J.–H. Seo et al., A simplified method of making flexible blue LEDs on a plastic substrate, *IEEE Photonics Journal*, Vol. 7, No. 2, pp. 1–7, April 2015, Art no. 8200207, 2015.

[55] Jaeyi Chun, Youngkyu Hwang, Yong-Seok Choi, Jae-Joon Kim, Tak Jeong, Jong Hyeob Baek, Heung Cho Ko, Seong-Ju Park, Laser lift-off transfer printing of patterned GaN

light-emitting diodes from sapphire to flexible substrates using a Cr/Au laser blocking layer, *Scripta Materialia*, Vol. 77, pp.13–16, 2014.

[56] Y.Y. Liu, X.Y. Wang, Y. Cao, X.D. Chen, S.F. Xie, X.J. Zheng, H.D. Zeng, A flexible blue light-emitting diode based on zno nanowire/polyaniline heterojunctions, *Journal of Nanomaterials*, Vol. 2013, p. 870254,2013.

[57] Y. Fujisaki, Y. Nakajima, T. Takei, H. Fukagawa, T. Yamamoto and H. Fujikake, Flexible active-matrix organic light-emitting diode display using air-stable organic semiconductor of dinaphtho[2, 3-b: 2',3'-f]thieno[3, 2-b]-thiophene, *IEEE Transactions on Electron Devices*, Vol. 59, No. 12, pp. 3442–3449, Dec. 2012.

[58] White, M., Kaltenbrunner, M., Głowacki, E. et al., Ultrathin, highly flexible and stretchable PLEDs. *Nature Photon* 7, 811–816 (2013).

[59] Y.–J. Wang, J.–G. Lu and H.–P. D. Shieh, Efficiency enhancement of organic light-emitting diodes on flexible substrate with patterned inverted conical structure, *IEEE Photonics Journal*, Vol. 8, No. 1, pp. 1–8, Feb. 2016.

[60] S.–M. Lee, J.H. Kwon, S. Kwon and K.C. Choi, A review of flexible oleds toward highly durable unusual displays, *IEEE Transactions on Electron Devices*, Vol. 64, No. 5, pp. 1922–1931, May 2017.

[61] Han, T.H., Park, M.H., Kwon, S.J. et al., Approaching ultimate flexible organic light-emitting diodes using a graphene anode. *NPG Asia Mater* 8, e303 (2016).

[62] Ok, K.H., Kim, J., Park, S.R. et al., Ultra-thin and smooth transparent electrode for flexible and leakage-free organic light-emitting diodes. *Sci Rep* 5, 9464 (2015).

[63] Su, Q., Zhang, H. and Chen, S. Flexible and tandem quantum-dot light-emitting diodes with individually addressable red/green/blue emission. *Npj Flex Electron* 5, 8 (2021).

[64] H.W. Hwang, S. Hong, S.S. Hwang, K.W. Kim, Y.M. Ha and H.J. Kim, Analysis of recoverable residual image characteristics of flexible organic light-emitting diode displays using polyimide substrates, *IEEE Electron Device Letters*, Vol. 40, No. 7, pp. 1108–1111, July 2019.

[65] Keum, C., Murawski, C., Archer, E. et al., A substrateless, flexible, and water-resistant organic light-emitting diode. *Nat Commun.*, Vol. 11, p. 6250 (2020).

[66] Yi-Ning Lai, Chih-Hao Chang, Pei-Chun Wang, Ying-Hao Chu, Highly efficient flexible organic light-emitting diodes based on a high-temperature durable mica substrate, *Organic Electronics*, Vol. 75, p. 105442, 2019.

[67] Ross E. Triambulo, Jin-Hoon Kim, Jin-Woo Park, Highly flexible organic light-emitting diodes on patterned Ag nanowire network transparent electrodes, *Organic Electronics*, Vol. 71, pp. 220–226, 2019.

[68] Ahreum Kim, Gunel Huseynova, Jonghee Lee, Jae-Hyun Lee, Enhancement of out-coupling efficiency of flexible organic light-emitting diodes fabricated on an MLA-patterned parylene substrate, *Organic Electronics*, Vol. 71, pp. 246–250, 2019.

[69] K.-J.Ko, S.-R.Shin, H.B.Lee, E.Jeong, Y.J.Yoo, H.M.Kim, Y.M.Song, J.Yun, J.-W.Kang, Fabrication of an oxide/metal/oxide structured electrode integrated with antireflective film to enhance performance in flexible organic light-emitting diodes, *Materials Today Energy*, Vol. 20, p. 100704, 2021.

[70] Anjali K. Sajeev, Nishkarsh Agarwal, Anjaly Soman, Shilpi Gupta, Monica Katiyar, A. Ajayaghosh, K.N. Narayanan Unni, Enhanced light extraction from organic light emitting diodes using a flexible polymer-nanoparticle scattering layer, *Organic Electronics*, Vol. 100, p. 106386, 2022.

[71] Eun Jeong Bae, Jong Woo Kim, Byeong-Kwon Ju, Dong-Hyun Baek, Young Wook Park, Flexible external light extraction in organic light-emitting diodes by porous PDMS film fabricated by high-pressure steam process, *Organic Electronics*, Vol. 108, p. 106575, 2022.

[72] Zhiyong Xiong, Wanlu Zhang, Zhongjie Cui, Shiliang Mei, Zhe Hu, Zhuoqi Wen, Haiyang He, Zhongtao Duan, Fengxian Xie, Ruiqian Guo, Eliminating hysteresis effects in flexible organic light-emitting diodes, *Organic Electronics*, Vol. 103, p. 106467, 2022.

[73] Lihui Liu, Shuling Li, Lei Wu, Dingfu Chen, Kun Cao, Yu Duan, Shufen Chen, Enhanced flexibility and stability of PEDOT:PSS electrodes through interfacial crosslinking for flexible organic light-emitting diodes, *Organic Electronics*, Vol. 89, p. 106047, 2021.

[74] Hui Du, Yangyang Guo, Dongyue Cui, Shuhong Li, Wenjun Wang, Yunlong Liu, Yicun Yao, Ling Zhao, Xiaochen Dong, Solution-processed PEDOT:PSS:GO/Ag NWs composite electrode for flexible organic light-emitting diodes, *SpectrochimicaActa Part A: Molecular and Biomolecular Spectroscopy*, Vol. 248, p. 119267, 2021.

8 Performance Enhancement of Light Emitting Radiating Dipoles (LERDs) Using Surface Plasmon-Coupled and Photonic Crystal-Coupled Emission Platforms

Seemesh Bhaskar and Sai Sathish Ramamurthy

CONTENTS

8.1 INTRODUCTION: BACKGROUND TO NANOPLASMONICS AND LERDS

Numerous interdisciplinary applications related to clinical diagnostics, microscopy, and sensing have been reported with the use of LERDs (such as flurophores and phosphorophores) [1–5]. These studies with LERDs are well supported with appropriate simulations and experimental data, thereby presenting a promising approach to quantify analytes at trace concentrations [6–9]. Additionally, LERDs are extensively explored in disciplines encompassing nonlinear optics, electro-optic catalysis as well as life sciences research [10,11]. In this chapter we refer to fluorescent emitters or emitter dipoles as LERDs to highlight their property to emit photons/light. Generally, understanding of the LERDs is based on their intrinsic ability to absorb and emit a particular wavelength of light [12,13]. Recently, the interaction of LERDs with microscale and nanoscale materials are progressively researched to understand the modulation in the emission intensity. It has been observed that the circumambient environment of the LERDs plays a crucial role in determining the emission intensity. Particularly, the presence of plasmonic or metallic nanoparticles (NPs) in the immediate vicinity of LERDs is reported to significantly alter their radiative properties [14–16]. Plasmonic NPs sustain localized surface plasmon resonance (LSPR) that generates an intense and concentrated electromagnetic (EM) field gradient around it. This attribute of plasmonic NPs has been extensively researched in different interdisciplinary fields including biosensing, solar cells, quantum optics, electrochemistry, microfluidics, antimicrobial, and anticancer applications to name a few [17–22]. In other words, the broad research utilizing the plasmonic response of NPs has grown into a fertile research arena termed 'nanoplasmonics'.

The close interaction of LERDs with the LSPR of plasmonic NPs drastically influence their lifetime and radiative decay rates, and this field has developed as a subfield of nanoplasmonics known as metal-enhanced fluorescence (MEF) [13,15]. While the overall extinction of the plasmonic NPs are determined by the contributing factors of scattering and absorption, the emission intensity of LERDs can be enhanced by increasing the scattering component of plasmonic NPs. In this regard, recently, plasmonic NPs of sharp nanogeometries and pointed tips (such as nanocubes, nanostars, nanotriangles) are increasingly explored [23–25]. The excitation and emission intensities of the LERDs are considerably improved in the presence of sharp-edged plasmonic NPs. Moreover, in a parallel phenomenon termed 'induced-plasmon effect', the excited state LERDs are reported to create/induce plasmons by directly interacting with the proximal plasmonic NPs [13]. This field of MEF has resulted in a paradigm shift concerning light-matter interactions of LERDs and plasmonic NPs and is widely used as promising technology for the detection of molecules and ions of interest at extremely low concentrations [26–29].

8.2 LERDS IN SURFACE PLASMON-COUPLED AND PHOTONIC-CRYSTAL COUPLED EMISSION

Although MEF has demonstrated better sensitivity and reliability as compared to conventional use of LERDs, certain drawbacks including inefficiency in the collection of emission, unavoidable quenching, and other experimental artifacts still remain the chief bottleneck [13,15,26,27]. This is majorly on account of the isotropic attribute of the emission pattern. In this context, metal thin films are studied in appropriate optical configurations for overcoming the above-mentioned drawback. While plasmonic NPs sustain LSPR, the metallic thin films support propagating surface plasmon polaritons (SPPs) in the metal-dielectric interface with an evanescent characteristic that is highly sensitive to slight changes in the refractive index of the dielectric medium. Recently, different plasmonic thin films (Ag, Au, Pt, Pd) are explored depending on the wavelength of interest with different thicknesses of dielectric nanolayer to comprehend the emission properties from LERDs that are coated on top of metal thin films [30]. Highly polarized (transverse magnetic) and sharply directional/channelized emission that supports more than 50% signal collection efficiency, emission (vis-a-vis < 1% in conventional LERDs) has been observed. Since the emission from the LERDs modulated based on the plasmonic coupling to the propagating SPPs of metal thin films, the resulting technology has been termed 'surface plasmon-coupled emission (SPCE)' [13,15]. On account of the unique characteristics of the SPCE platform, such as high collection, low-background noise, and excellent spectral resolution, this technology pioneered by Lakowicz et al., has established a strong foothold in the broad arena of spectro-plasmonics [15]. The SPCE platform has been extensively explored to comprehend different processes and phenomena at functional nanointerfaces including kinetics of adsorption-desorption, multi-analyte detection, smartphone-based biosensing, molecular beacon effects, novel photo-plasmonic waveguides, excitation–emission synchronization, energy transfer in monomer-higher order aggregates, polymer light emitting diodes to name a few [31–33]. These studies have resulted in new physicochemical outcomes as well as real time applications in on-field biosensing approaches.

Despite several advantages of the SPCE platform, the intrinsic parasitic Ohmic losses in metal-based detection platforms always remain a hindrance to attain expected sensitivity. To address this caveat of losses in metal-based sensor platforms, the STAR laboratory in the year 2020 demonstrated the utility of one-dimensional photonic crystal (1DPhC) as an essential lossless analog to SPCE substrate with the development of photonic crystal-coupled emission (PCCE) platform [4]. This is made up of quarter wave plate thick alternating nanolayers of high refractive index (HRI) and low refractive index (LRI) nanolayers fabricated in such a way so as to obtain a photonic band gap (PBG). The PBG is the region where a certain set of wavelengths are not allowed by the substrate for normal incidence of light. The elaborate fabrication details of 1DPhC along with cross-sectional imaging and experimental reflectance data is presented in our recent reports [4,6,8,9]. Generally, Kretschmann–Raether (KR) and reverse Kretschmann (RK) prism-coupled optical configurations are adopted for these experiments. In the SPCE and PCCE experimentation described in this chapter the authors have adopted the RK optical configuration (Figure 8.1). Here the laser light irradiates the engineered SPCE or PCCE substrates directly from the free space (FS) side (and not from the curved surface of the prism side). Appropriate filters and polarizers are used to collect the emission [34,35]. Both the free space (FS) and coupled emission (from SPCE and PCCE platforms) were recorded using a USB4000

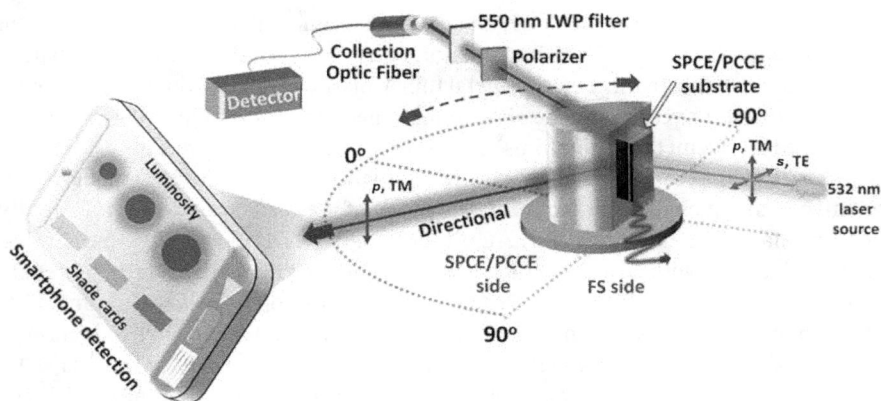

FIGURE 8.1 General RK optical configuration adopted in SPCE and PCCE experimentation. The smartphone-based detection as well as the conventional Ocean optics based detection is also presented. Reprinted with permission from Reference [12], Copyright 2021, American Chemical Society.

fiber optic spectrometer connected to Ocean Optics SpectraSuite software. Polyvinyl alcohol (PVA) has been used as the dielectric nanolayer for doping additional nanomaterial in order to maintain generality across all the experiments. While the 50 nm Ag thin film vapor deposited on Pyrex glass slide constitutes the SPCE platform, the PCCE platform is made up of 1DPhC substrate. Further, while the surface plasmons couple with LERDs in the case of the SPCE platform, the Bloch surface waves (BSWs) and internal optical modes (IOMs) of PCCE substrate couple with the LERDs in the case of PCCE platform. On account of the lossless attributes of the dielectric-based PCCE platform, the sensitivity is drastically enhanced as compared to the metal-based SPCE platform.

8.3 NANOENGINEERING STRATEGIES FOR AUGMENTING THE PERFORMANCE OF LERDS

The SPCE and PCCE platforms are recently being explored with pragmatic nanoengineering methodologies to further improve the performance of the LERDs. This is achieved by integrating the well-established MEF technique with the SPCE platform in difference functional nanointerfaces such as (i) spacer; (ii) cavity; and (iii) extended (ext.) cavity [28,29]. Subwavelength localization of the incident EM radiation at nanoscale domains (between the plasmonic NPs and metal thin film) significantly influences the local density of states (LDOS) of LERDs. Depending upon the nanointerface adopted, the plasmonic coupling efficiency is modulated as the hotspot intensity is altered [36,37].

The nanointerfaces are conceptually depicted in Figure 8.2 and the elaborate details are presented in earlier publications [35–38]. Only a brief description of the nanointerface architectural design is presented in this section to familiarize the readers. The spacer nanointerface is an architecture wherein the NPs doped in a polymer matrix are spin coated as the first film layer followed by polymer matrix doped with LERDs. The emission from the LERDs is noticeably reformed depending on the nature of the spacer layer in this case. The NPs and the LERDs are mixed prior to spin coating as a single layer (with the help of polymer matrix) in the cavity nanointerface. Consequently, LERDs are sandwiched in the hotspot region in the nanogap between the NPs and the metal thin film or 1DPhC. Depending upon the tradeoff between the quenching effects (as the NPs and LERDs are admixed) and the hotspot intensity the emission intensity is either dwindled or enhanced. The ext. cavity interface in the reverse format of spacer and presented the best of both the advantages of the spacer and cavity nanointerface. Here the LERDs are coated first, followed by the NPs doped in polymer matrix. As the LERDs are not admixed with the NPs, the surface induced quenching can be avoided in addition to presenting a spacer nanolayer. By and large these nanointerfaces have been adopted for detection of ions and molecules, as described in the subsequent sections.

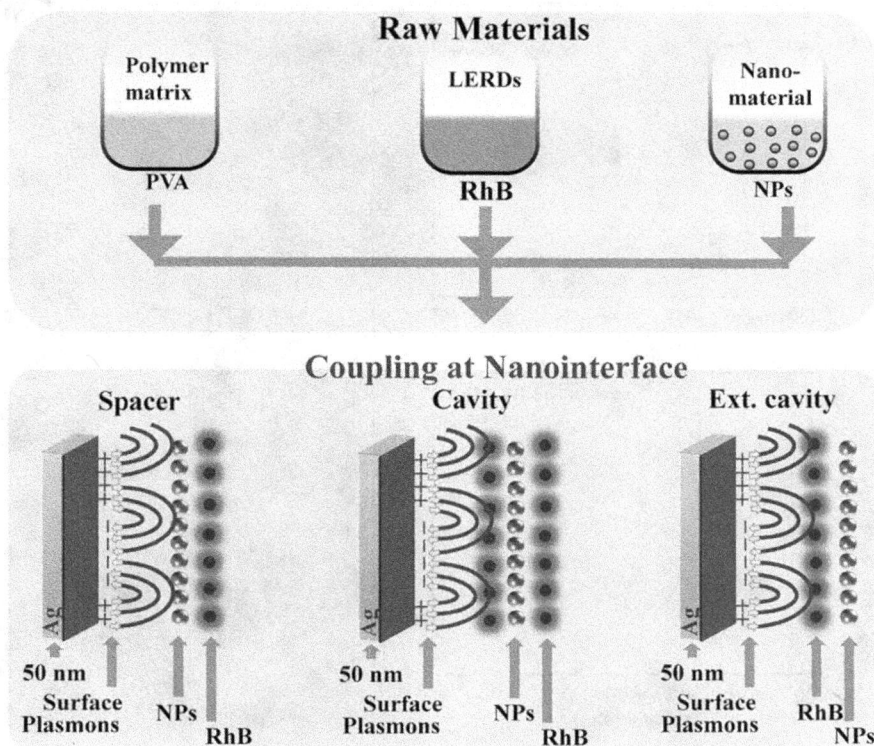

FIGURE 8.2 Conceptual schematic of nanointerfaces: spacer, cavity, and ext. cavity is showing along with the starting or raw materials used for the fabrication of these nanointerfaces. Reprinted with permission from Reference [12], Copyright 2021, American Chemical Society.

8.4 NANOSTRUCTURE ACTIVITY RELATIONSHIP TOWARDS DEQUENCHING LERDS

Although different plasmonic NPs have been exponentially investigated in the SPCE platform to understand the photo-plasmonic response and coupling efficiency, the utility of AuNPs have been minimal. This is on account of the well-known 'zone of inactivity', which is a region of < 5 nm distance between the AuNPs and the LERDs (typically rhodamine moieties) [25,28,38]. On account of the enormous interband losses exhibited by the AuNPs, the coupling occurs via the non-radiative higher order modes that does not yield effective emission due to energy dissipation. Consequently, in such close proximity of LERDs to AuNPs quenching has been reported to dominate the effective emission augmentation.

In order to address this drawback, different techniques have been developed including development of core-shell architectures, metal-dielectric nanointerfaces, and sharp-edged nanostructures [25,28]. While each of the methodologies present

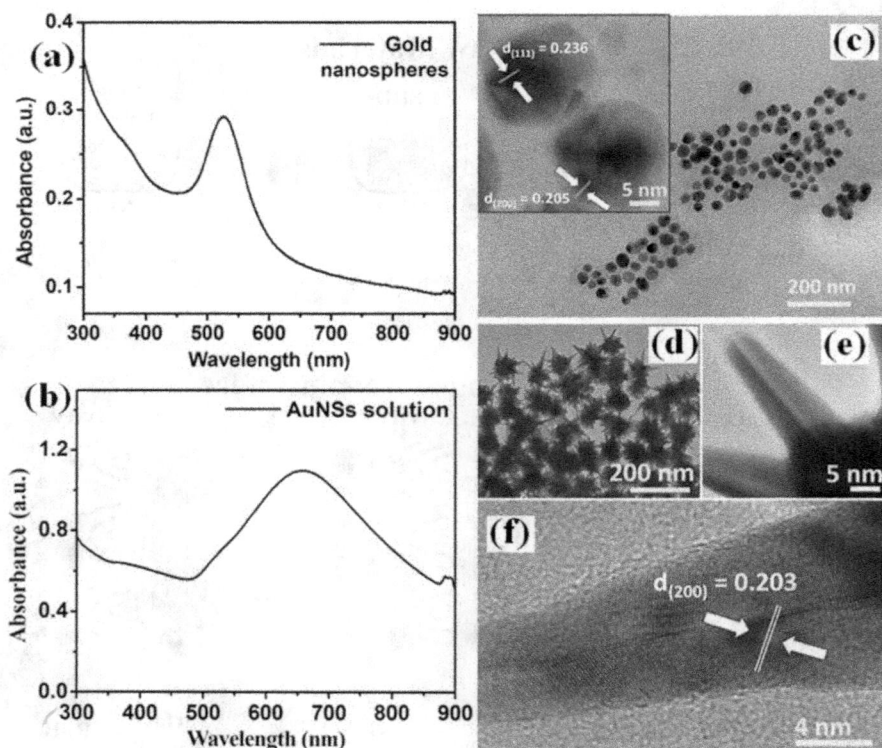

FIGURE 8.3 UV–visible absorbance spectra of (a) AuNPs and (b) AuNSs. (c) TEM image of AuNPs and HRTEM image of AuNPs shown as an inset indicating the d-spacing characteristic to gold. (d), (e) TEM images of synthesized AuNSs. (f)HRTEM image of AuNSs presenting the d-spacing and Miller indices characteristic to gold. Reprinted with permission from Reference [25], Copyright 2020, Elsevier.

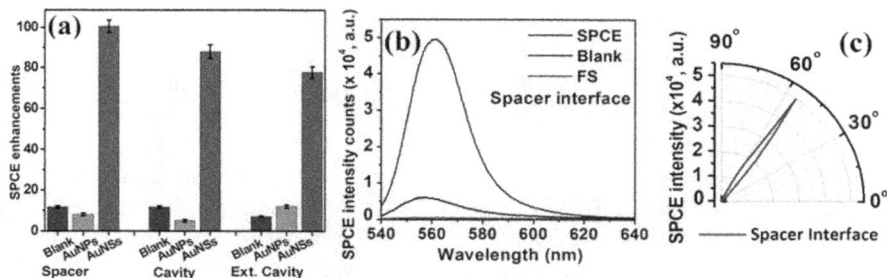

FIGURE 8.4 (a) Comparative assessment of SPCE emission enhancements for AuNPs and AuNSs along with respective blanks in spacer, cavity, and ext. cavity interfaces. (b) SPCE and FS emission intensity obtained with AuNSs along with blank in spacer interface. (c) Angularity plot for SPCE with AuNSs in spacer interface. (b) and (c) are chosen for representation because spacer interface presented maximum emission enhancements. Reprinted with permission from Reference [25], Copyright 2020, Elsevier.

unique advantages and disadvantages, it is instructive to discuss in particular about the ability of structural morphology of nanomaterials to have an influence on the performance of LERDs. This section is devoted to discuss the aspects of structure-activity relationship of AuNPs and Au nanostars (NS) towards their interaction with adjacently located LERDs. The AuNPs and the AuNS synthesis are presented in detail in the associated works [25]. The UV-visible absorbance spectra, TEM and HRTEM images are presented in Figure 8.3. The TEM images clearly depicts the sharp-edged nanostar morphology with the characteristic lattice fringes shown in the HRTEM images corresponding to that of crystalline Au. The SPCE results obtained with these materials are shown in Figure 8.4. It is clearly observed that across all the three nanointerfaces the SPCE enhancements decreased for AuNPs (due to quenching) as compared to blank samples (without any NPs).

Further, SPCE enhancements observed for AuNSs indicate that the sharp-edged NSs not only dequenched the emission signal but also resulted in augmented enhancements from LERDs. This is primarily on account of the tip plasmons sustained by the nanotips of sharp protrusions, as well due to the lightning rod effect supported by such elongated nanospikes [25]. In light of these observations, we conclude that the structural morphology plays a significant role in determining the response of the LERDs present in the proximal distance from Au nanomaterials. For the sample yielding the highest SPCE enhancement, the SPCE intensity spectra and the sharply directional angularity plot are shown in Figures 8.4b and 8.4c, respectively.

8.5 METAL, DIELECTRIC, AND METAL-DIELECTRIC NANOHYBRIDS FOR BIOSENSING USING LERDS IN THE SPCE PLATFORM

In this section, we discuss the four major approaches adopted to enhance the performance of LERDs for biosensing applications. In the year 2004, Chowdhury and co-workers demonstrated an efficient nanoengineering technique wherein the AgNPs were explored as a spacer in the SPCE platform [39]. This yielded 60-fold SPCE enhancements, as compared to 10-fold enhancements observed in blank samples without NPs. After this initial discovery, different nanomaterials are explored in the SPCE platform and can be broadly categorized into four domains: (i) metal nanoassembly (sorets); (ii) metal-dielectric nanoassembly (Au-SiO$_2$ nanohybrids); (iii) dielectric NPs(Nd$_2$O$_3$, TiC, TiN, TiCN NPs); (iv) anisotropic metallic NPs (Ag and Au). This has been systematically performed in order to overcome the Ohmic losses and surface induced quenching effects observed in metallic plasmons [40,41]. While pristine plasmonic NPs sustain LSPR, the nanoassemblies sustain nanovoids and nanocavities while studied in the SPCE platform (i.e., in nanoparticle-on-a-mirror configuration). In this context, soret colloids, which are precise nanoassemblies have been explored [7,8,9,16]. In brief, soretnanoassemblies are synthesized using the well-known methodology of adiabatic cooling technique by subjecting the homogenous NPs' solution to temperature gradient [42,43]. From extensive plasmonic analysis it has been observed that~100-fold SPCE enhancements can be achieved using soretnanoassemblies in spacer nanointerface (as compared to 60-fold observed with pristine NPs) [7,9,16,37].

This high enhancement was utilized for sensing biologically relevant glutathione molecule [7]. The close interaction among the ARS-Cu^{2+}-GSH ensemble was utilized for development of the GSH sensor. High fluorescence emission from 0.5 μM concentration of the ARS molecule can be obtained in a pH 5.5 Britton– Robinson (BR) buffer (composition: 0.04 M phosphoric acid, 0.04 M boric acid (BA), 0.04 M acetic acid and 0.2 M sodium hydroxide) on account of the formation of fluorescent ARS-BA complex. On account of the high collection efficiency of SPCE platform ~7-fold enhancements were observed for ARS solution in BR buffer. With the addition of AgSC, ~97-fold SPCE enhancements observed due the inter-plasmon coupling between the AgSC and Ag thin film (comparatively, pristine AgNPs yielded ~45-fold) [7]. As observed in Figure 5a, with the addition of Cu^{2+} ions (from 0 nM to 520 nM in 40 nM concentration intervals) to the ARS-BA complex, quenching was observed due to the preferential binding of ARS to Cu^{2+} ions as compared to BA. Further, good correspondence was observed between the SPCE enhancements and luminosity (Figure 8.5b). The gray-scale shade cards associated with the respective luminosity values are shown vertically on the right.

Further, the GSH addition to the above ARS−Cu^{2+} complex, resulted in effective dequenching of emission (Figure 8.5c), which is also called as 'turn-on' fluorescence. The respective SPCE enhancements along with the luminosity values are shown in Figure 8.5d along with the shade cards shown in the right. On account of the simple, frugal, rapid quenching-dequenching technique adopted for detection of GSH at femtomolar sensitivity, one can envisage the ability of such soretnanoassemblies to assist various biomolecule sensing. Further details of the highlights of GSH sensing are detailed in the associated publication [7]. While this serves as a representative example for the ability of nanoassemblies to achieve better sensitivity in biosensing, it must not go without mention that nanoassemblies made up of pristine plasmonic NPs still suffer from Ohmic losses.

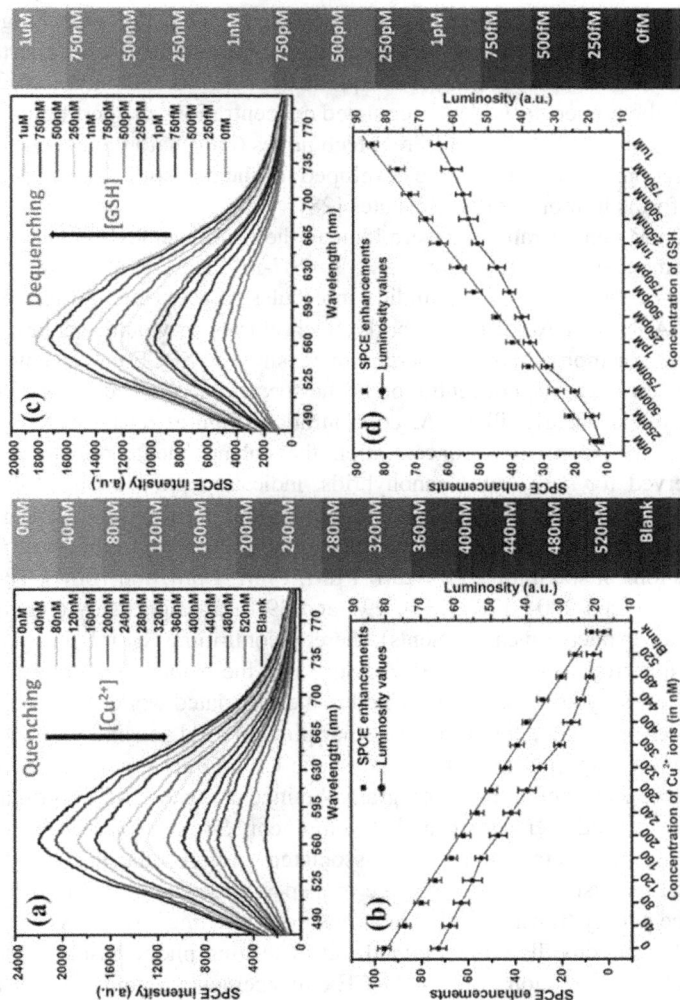

FIGURE 8.5 (a) Quenching in the SPCE spectra (of ARS) upon addition of higher concentration of Cu²⁺ions (b) SPCE enhancements and associated luminosity values corresponding to the Figure 5a. (c) Dequenching in the SPCE spectrum of mixture of ARS + Cu²⁺ (520 nM) with increasing GSH concentration. (d) SPCE enhancements and associated luminosity values corresponding to the Figure 8.5c. AgSC are used for the sensing work in spacer nanointerface. Reprinted with permission from Reference [7], Copyright 2020, American Chemical Society.

In an attempt to address this lacuna, in another work we demonstrated the robustness of metal-dielectric Au-SiO$_2$ nanohybrids (abbreviated as AuSil) to not only overcome the quenching limitation observed with AuNPs, but also yield augmented fluorescence enhancements. In this novel technique, AuNPs were decorated differently on silica NPs to obtain tunable SPCE enhancements [28]. The AuSil sample yielding the highest SPCE enhancements was utilized for sensing the spermidine molecule on account of the close interaction of AuSil with spermidine [28]. This sensing was performed in extended cavity nanointerface to avoid the quenching of LERDs observed in cavity and spacer nanointerfaces. Spermidine is an important bioamine present in humans, plants, as well as aquatic life systems and plays a crucial role in maintaining the homeostasis. As unregulated concentrations of spermidine are associated with several disease states, different techniques (chromatography, electrochemical, spectroscopic to name a few) are developed for the detection of spermidine in a variety of lifeforms to monitor disease states [28].

Addition of 100 μM concentration of spermidine to the AuSil nanohybrid increased the SPCE enhancement from 88-fold (for AuSil) to 207-fold as seen in Figure 8.6a. This is on account of the ability of spermidine molecules to assist the clustering of adjacently present AuNPs of AuSil nanohybrids. Such clustering would increase the number of available plasmonic hotspots thereby increasing the SPCE enhancements. Further, upon the decrease in the concentration of the spermidine molecule, the SPCE enhancements decreased linearly till 10 fM concentration (Figures 8.6a, 8.6b). Upon further decrease in the spermidine concentration the enhancements remained the same as that observed for bare AuSil nanohybrids, indicating the reliability of the sensor only up to femtomolar concentrations. Furthermore, in order to validate the reliability of the developed sensor platform, spiking studies were performed with five different concentrations of spermidine (100 μM, 1 μM, 1 nM, 1 pM, and 10fM). These five concentrations yielded 200-, 178-, 135-, 99-, and 89-fold, respectively (standard deviation of ±5% for triplicate measurements), thereby confirming the reliability and reproducibility of the proposed sensor [28]. Additionally, the elaborate details of the selectivity and the recovery analysis are presented in the associated work [28].

In order to present portable, cost-effective smartphone-based sensing of spermidine, the emission was captured and processed in the ColorGrab app (obtained via the Google Play Store). The chromaticity diagram obtained using the CIE coordinates are being extensively used in research and development. Some examples include the studies involving the photoluminescence associated with correlated color temperature (CCT), pH analysis with color change in a specific reaction, color changes observed during the energy transfer in chemical reactions to name a few [28]. In this background, the chromaticity diagram obtained using the smartphone based analysis presented an excellent correlation with the SPCE enhancements obtained using the expensive detector system. The consistent red-shift in the emission spectral profile of RhB observed with changing concentration of spermidine is the reason for the observed consistent shift in the associated chromaticity diagram.

In the earlier sections we systematically discussed the utility of plasmonic/metallic nanoassemblies (AuSil) and metal-dielectric (AuSil) nanoassemblies to augment the performance of LERDs in the SPCE platform. It was observed that much higher yield

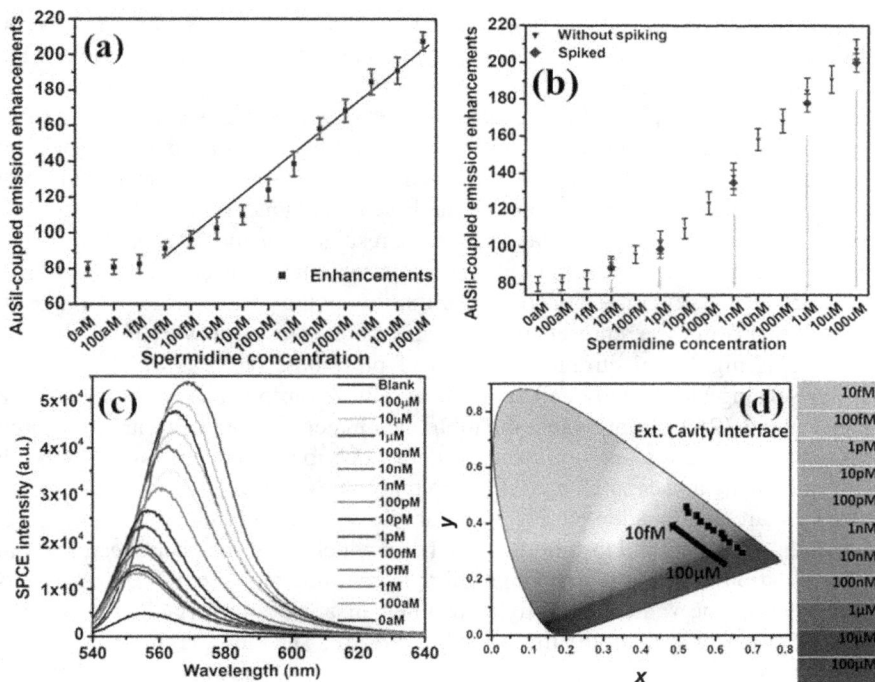

FIGURE 8.6 (a) SPCE enhancements obtained for different concentrations of spermidine. (b) SPCE enhancements obtained with different spermidine concentration with (red) and without (blue) spiking spermidine. Green color connecting lines are shown for spiked samples for easy visualization. (c) SPCE spectra observed for different concentrations of spermidine. One can observe the consistent shift in the spectral profile. (d) CIE chromaticity plot indicating the shift in the color for different concentrations of spermidine shown with respective shade cards in the right. Reprinted with permission from Reference [28], Copyright 2020, American Chemical Society.

can be obtained with the incorporation of dielectric NPs along with the metallic NPs as in the case of AuSil. The ability of dielectric NPs to present optical confinement without exhibiting losses has been deliberated in different research publications for a variety of applications [40,41]. In this direction, we intended to explore the utility of pure/pristine dielectric NPs of high refractive index (HRI). In this context, HRI dielectric Nd_2O_3 nanorods (NRs) were studied in the SPCE platform. While the plasmonic Ag soretnanoassemblies yielded 104-fold enhancement, the Nd_2O_3 NRs yielded 118-fold enhancement of emission from LERDs in the SPCE platform [23]. This is worthy of emphasis as it is possible to obtain higher performance with the use of anisotropic HRI dielectric NRs. This high enhancement from LERDs was utilized for detection of environmentally hazardous tannic acid molecule. The close interaction of Nd_2O_3 NRs with the tannic acid aids in declumping of the Nd_2O_3 NRs' agglomerates [23]. Consequently, a greater number of NRs are released into the effective area, increasing the number of photo-plasmonic hotspots. The lightning-rod effect supported by the rod-shaped morphology of Nd_2O_3 NRs is the primary reason for observation of high enhancement in the fluorescence intensity [8,9,16,23]. The details of need for detection of tannic acid in drinking water samples, relevance of sensing, extensive analysis of results are detailed in the associated work [23]. The picomolar limit of detection of tannic acid observed from high SPCE enhancements has also been analyzed using a simple, easy-to-use, portable smartphone-based technique, with associated luminosity values and excellent correlation with the results from conventional detectors.

Furthermore, towards the culmination of this section we introduce the readers to the importance of structural anisotropy for improving the performance of LERDs in the SPCE platform. As a representative example a recent research work is highlighted that presents highly anisotropic structures of plasmonic AgNPs, which has been used for sensing biologically and environmentally hazardous mercury ions at trace concentrations [24]. Of late, biocompatible nanomaterials obtained via bioinspired sustainable routes are gaining increased interest due to the concern associated with harmful reducing and capping agents [5]. In this context, recently, bio-nano-inspired techniques are being adopted for nanoparticle synthesis [44,45]. The Lycoat® polymer is a biocompatible one that has been widely used in pharmaceuticals in the formulation of tablets and drugs. Simple mixture of Lycoat® and Ag^+ ions and subjecting the same to UV-light, resulted in anisotropic AgNPs, of tunable mohphology depending on the time of UV exposure. The unique functional groups present in Lycoat® acts as capping and reducing agents. Thus obtained AgNPs with cubic morphology yielded >900-fold enhancement in the emission from LERDs in ext. cavity nanointerface.

This high enhancement was utilized for sensing of mercury ions at extremely low concentrations (attomolar LOD) [24]. The close interaction of sharp-edged AgNPs (nanocubes) and Hg^{2+} ions results in disintegration of AgNPs into Ag^+ ions as shown in Figure 8.7a. The disappearance of the characteristic LSPR of AgNPs in the UV-vis absorbance spectra with increasing concentration of Hg^{2+} ions is presented in Figure 8.7b. A visual change in the color of the solution from characteristic yellow to colorless is observed with increasing concentration of Hg^{2+} ions (Figure 8.7c). As observed in Figure 8.7d, with increasing concentration of Hg^{2+} ions the SPCE enhancements decreased. This is because of the reduction in the effective hotspot density due to disintegration of AgNPs to Ag^+ ions. As the hotspots are reduced the plasmonic coupling of AgNPs with metal thin film is reduced, thereby resulting in dwindled SPCE enhancements. As observed in Figure 8.7d, an excellent correlation has been observed between the SPCE enhancements (obtained using spectrophotometer) and the luminosity values (obtained using smartphone-based detection platform). The shade cards corresponding to the luminosity values are presented in Figure 8.7e. The smartphone-based analysis aids in easy correlation of the data obtained in monitoring concentration of Hg^{2+} ions, especially in resource limited settings. Hence, we conclude that anisotropic nanomaterials with unique nanogeometries can perform significantly toward enhancing the sensitivity and overall emission from LERDs. It can be emphasized that such anisotropic geometries can be obtained via frugal nanoengineering techniques without the use of harmful capping and reducing agents.

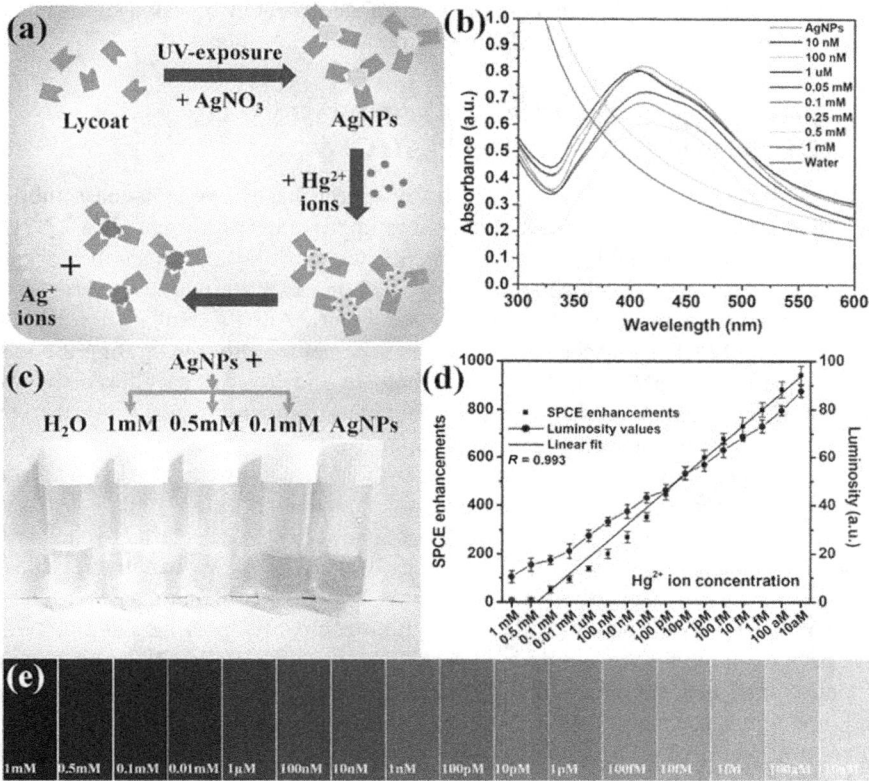

FIGURE 8.7 (a) Conceptual illustration of the steps involved in the disintegration of AgNPs to Ag⁺ ions upon the addition of Hg²⁺ ions. (b) UV–vis absorbance spectra observed with different concentration of Hg²⁺ ions addition. (c) Visual change in the color of the solution of AgNPs observed with addition of different concentrations of Hg²⁺ ions. (d) Correlation between the SPCE enhancements (gathered via spectrophotometer) and the luminosity data (gathered via smartphone). (e) Shade cards corresponding to different luminosity values shown in Figure 8.7d. Reprinted with permission from Reference [24], Copyright 2021, American Chemical Society.

In conclusion, in this section we have deliberated upon three major routes that can be employed for enhancing the performance of LERDs in the SPCE platform. The journey from metallic nanoassemblies (sorets) to metal-dielectric nanoassemblies (AuSil) to pristine anisotropic HRI dielectric NPs highlights the need for overcoming the Ohmic losses in metals using different novel methodologies. In addition to this, we also discuss the utility of anisotropy in structural morphology not only in pristine dielectric structures, but also pristine plasmonic/metallic nanostructures. To put it in a nut shell, we believe that further research in this direction would establish the appropriate balance between metal, dielectric, and their assembly and/or structural anisotropy needed to obtain the unprecedented enhancement in emission signal intensity from LERDs coated over the SPCE substrates.

8.6   BIMETALLIC NANOHYBRIDS FOR BIOSENSING USING LERDS IN THE SPCE PLATFORM

In the earlier section, we discussed certain key approaches where in nanoassemblies and metal-dielectric nanohybrids are utilized to improve the performance of LERDs in the SPCE platform. In this penultimate section we throw light upon another unique approach that has been developed in the year 2020 onwards, for not only overcoming the unavoidable quenching effects encountered with AuNPs, but also to realize unaccustomed SPCE enhancements. This has been achieved by integrating the best of the properties of AgNPs and AuNPs in a single hybrid. To crown it all, such hybrid integration has been achieved along with modulation in structural anisotropy, thereby presenting new channels for hotspot generation.

By and large, from the perspective of hetero-metallic nanostructures two synthesis techniques are reported, namely; (i) AgPtsoretnanoassemblies [37] and (ii) AgAu nanohybrids [12,14]. While the AgPtsoretnanoassmblies have been effective towards picomolar sensing of copper ions in drinking water, the AgAu nanohybrids have presented much better sensitivity. In this context, in order to obtain tunable structural morphologies of AgAu nanohybrids, different biocompatible polymers and proteins have been explored for the synthesis of AgAu nanohybrids. While silk worm protein yielded unique nanocuboidal structures (aiding attomolar sensitivity of zinc ions), the biopolymers such as soluplus® [12], kollidon® [29], and gelucire® [14] yielded different nanostructural morphologies and was utilized for the detection of different analytes at extremely low concentrations. The discussion in this section is majorly focused on the utility of biopolymers to generate unique nanohybrids that in turn yield higher SPCE enhancements and better sensitivity.

To begin with, the AgAu nanohybrids obtained using the three polymers (i) soluplus®; (ii) kollidon®; and (iii) gelucire® were synthesized using a general protocol. In this technique, the Ag^+ ions (1 mM) and Au^{3+} ions (1 mM) were admixed with 1 weight percent of the biopolymer and subjected to UV-light irradiation for different time intervals. The AgAu nanohybrids thus obtained were extensively characterized using UV-vis-NIR absorbance, DLS, SEM, and EDAX analysis and the details are presented in earlier publications [12,14,29]. The sample yielding the highest SPCE enhancements with exceptional EM hotspot density was further considered for sensing application. It should not go without mention that the AgAu nanohybrids were considered over other bimetallic combinations due to two major reasons: (i) high chemical stability of AuNPs as compared to AgNPs and (ii) low Ohmic losses of AgNPs as compared to AuNPs. Upon combining both the Ag and Au properties in a single hybrid, it was observed that the best properties of both the systems can be inherited into a single integrated system. Moreover, recent reports highlight the concept of plasmon passages in AuAgAu nanohybrids where the lossy property of AgNP sandwiched between the two chemically stable AuNPs can be completely overcome due to the nanostructural design [5,12,14,29]. Consequently, loss-less plasmonic passages can be realized using AuAgAu nanohybrids. With this background, we have intended to explore the utility of AgAu nanohybrids in SPCE platform to comprehend the ability of these structures towards fluorescence enhancements.

Soluplus®-mediated AgAu nanohybrids yielded 1200-fold SPCE enhancements. This has been utilized for sensing the SPCE reporter molecule RhB in ext. cavity nanointerface. It is important to note that RhB is a representative molecule for the general class of LERDs that include other fluorescent dyes. Additionally, it is important to develop an efficient technique for detection of RhB at low concentrations on account of their wide spread use in industries related to pharmaceuticals, ceramics leather, plastic, textiles to name a few [12]. Moreover, the altered concentrations of RhB in water bodies has been reported to cause serious health risks, including genotoxicity, carcinogenicity, and neurotoxicity thereby classifying such materials as xenobiotics. In spite of its hazardous characteristics, these dye systems are extensively used in research and industries due to their high quantum yield as well as effective coloring properties. In this background, different techniques are developed for sensing RhB samples at trace concentrations. As shown in Figure 8.8, two linear ranges were observed for sensing RhB (1 mM to 0.1 nM and 0.1 nM to 0.01 aM). With decreasing concentration of RhB the SPCE enhancements consistently decreased with a limit of detection of 0.01 aM. The augmented sensitivity is attributed to the high 1200-fold enhancements obtained using AgAu nanohybrids [12] (Figure 8.8).

In another approach, AgAu nanohybrids with unique nanocubic morphology is reported with the use of Gelucire® as reducing and capping agent. Gelucire® contain PEG esters and has the chemical name as Stearoyl macrogol-32 glycerides with the ability to generate microemulsions in contact with aqueous media [14,35]. It is a biocompatible solubility enhancer being extensively used in the pharmaceutical industry for stabilization of topical formulations. Although the utility of Gelucire® has been explored in the above mentioned applications, their importance in nanotechnology and plasmonics is seldom reported. The AgAu nanohybrids synthesized using a simple UV-exposure technique yielded cubic morphologies that sustain intense EM field intensity while studied in the SPCE platform. The nanocorners and the nanoedges of the nanocubes assist in significant localization of electron cloud density generating plasmonic hotspots [3–6]. As a result of this, the nanocubes synthesized via this approach yielded >1000-fold SPCE enhancements. The observed high SPCE signal intensity was utilized for smartphone-based attomolar detection of biologically relevant cysteine molecule as shown in Figure 8.9.

A simple chemical strategy involving the interaction between ARS (Alizarin Red-S)–Cu^{2+} ions and cysteine was utilized for the detection of cysteine molecules in the SPCE framework. As described in the earlier sections, the ARS molecules fluoresce while taken in the presence of BR (Britton-Robinson) buffer upon 400 nm absorbance [14]. Since we are using the 532 nm source, the quantum yield is significantly lower for ARS solution. In this context, we add Cu^{2+} ions to the ARS solution, as it results in the complexation, which in turn shifts the absorbance from 400 nm to 510 nm, which is effectively closer to that of the laser source used in the experiments. Consequently, we observe enhanced SPCE signal intensity upon 532 nm excitation. While cysteine

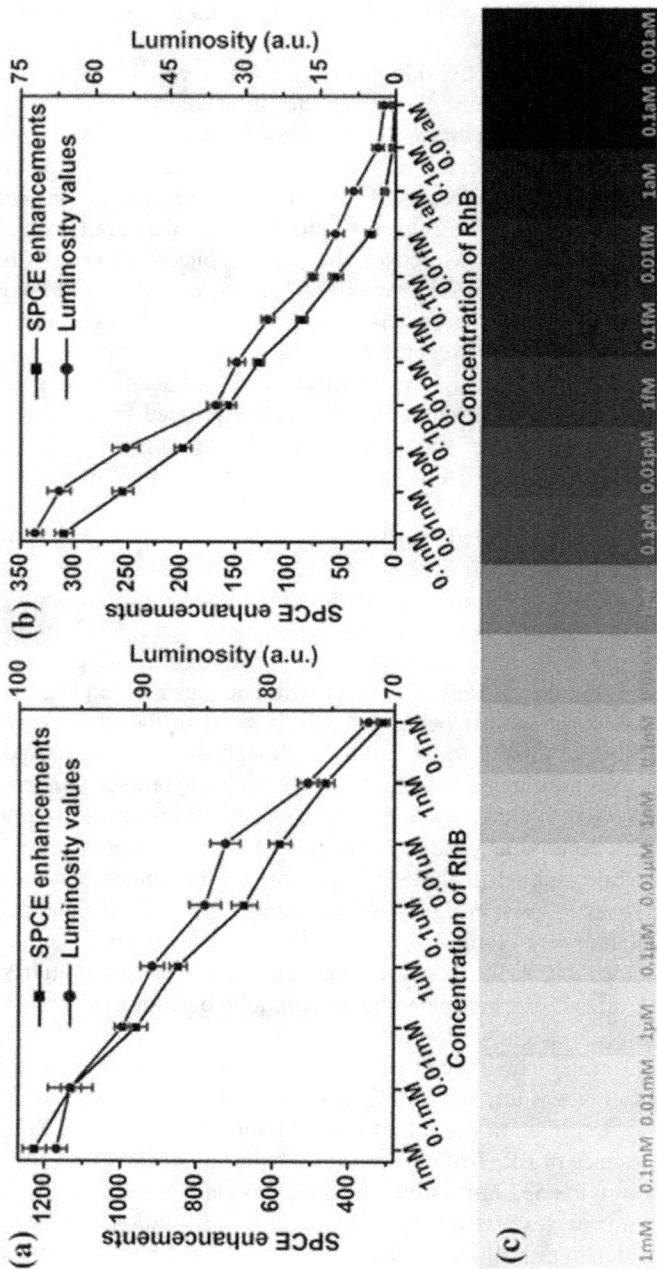

FIGURE 8.8 SPCE enhancements (left y-axis) and luminosity (right y-axis) for concentration range (a) 1 mM to 0.1 nM and (b) 0.1 nM to 0.01 aM. The AgAu nanohybrid sample obtained using soluplus as the reducing and capping agent with 30 min UV exposure was studied in the ext. cavity nanointerface for carrying out all the sensing experiments. (c) The gray-scale shade cards corresponding to the luminosity values presented in Figure 8a and 8b. Reprinted with permission from Reference [12], Copyright 2021, American Chemical Society.

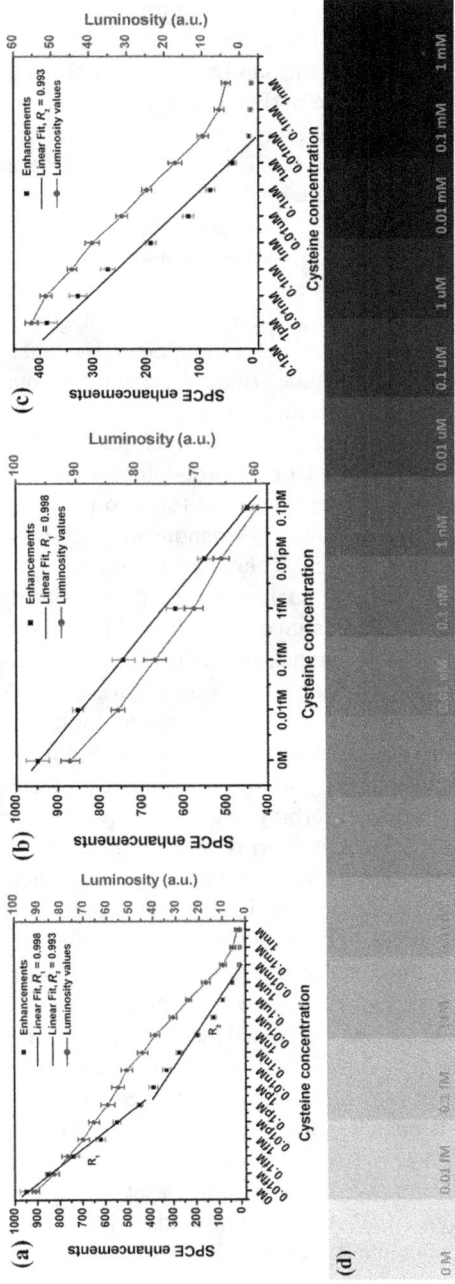

FIGURE 8.9 Sensing performed using Gelucire® based AgAu nanohybrids. Overlap of SPCE enhancements (obtained via conventional detector) and the luminosity (obtained via smartphone detector) given in the left and right y-axis, respectively, for (a) complete range of cysteine concentration, (b) in the concentration range of 0.01 fM to 1 mM and (c) in the concentration range of 1pM to 1 mM. (d) The gray-scale shade cards corresponding to the luminosity values. Reprinted with permission from Reference [14], Copyright 2022, Elsevier.

molecules are added to this mixture, competitive interaction of the same with ARS results in the unbinding of the Cu^{2+} and ARS ensemble. Consequently, the ARS returns to the native form with the absorbance at 400 nm, thereby presenting reduced SPCE signal intensity. In this process, increasing cysteine concentration results in a decrease in the SPCE enhancements and the results are presented in Figure 8.9. Once again, SPCE experimentation were performed by replacing the expensive conventional detectors with smartphone camera as the detector and the thus obtained luminosity values are plotted against the SPCE enhancements (Figure 8.9a–c). The gray-scale shade cards are shown in Figure 8.9d, and the respective luminosity values show an excellent correlation with the SPCE enhancements.

Research studies have demonstrated inflammation and physiological ailments on account of unregulated concentrations of amino acids. As a result of this, the altered concentrations of amino acids are utilized as biomarkers in diagnostics and therapy [29]. In this background, sensing and qualitative and quantitative analysis of tyrosine has been the topic of interest in fields concerning food and pharma sectors. Early detection of tyrosine assists in understanding the occurrence of related diseases in early stages. Although there are different strategies developed for the early detection of tyrosine at low concentrations, the existing methodologies are generally tedious, demanding large sample volumes and expert labor force to perform the associated analysis [29]. Consequently, in the past few years, nanotechnology is being synergized with novel plasmonic and electrochemical techniques for realizing ultra-high sensitivity of tyrosine molecules. In spite of such attempts, the sensitivity achieved has been in the femtomolar concentration range [29]. In light of these observations, we have utilized AgAu nanohybrids synthesized using Kollidon® for lowering the limit of detection to attomolar levels as presented in Figure 8.10. The sensing has been performed using a cost-effective smartphone-based tool and the interpretation is provided using the chromaticity diagram shown in Figure 8c. The sensing is accomplished using the simple strategy where the tyrosine molecules are reported to result in high quenching of green absorbing dye molecules, with the formation of ground-state complexes. The Ag/Au nanohybrids obtained using Kollidon® as reducing and capping agent yielded sharp nanorod architectures [29]. Such nanomaterials are ideal candidates for plasmonic coupling-based experiments as high EM field intensity can be observed due to the well-known lighting rod effect. The high >1100-fold SPCE enhancements obtained using AgAu nanorods were utilized for detection of tyrosine upon its interaction with RhB dye molecules. As observed from Figure 8.10, the SPCE enhancements decreased with increasing concentrations of tyrosine. Although significant quenching was observed with tyrosine molecules, with as high as 0.1 mM concentration of tyrosine, ~200-fold SPCE enhancements were still observed. This may be attributed to the ability of the AgAu nanorods to overcome the conventional quenching observed with tyrosine, on account of the abundant EM hotspots supported by them. Similarly, different strategies were adopted for the development of hetermoletallicAgAu and AgPt nanohybrids with several other unique geometries for sensing environmentally, chemically, and biologically relevant ions and molecules at extremely low concentrations. Interestingly, excellent correlation

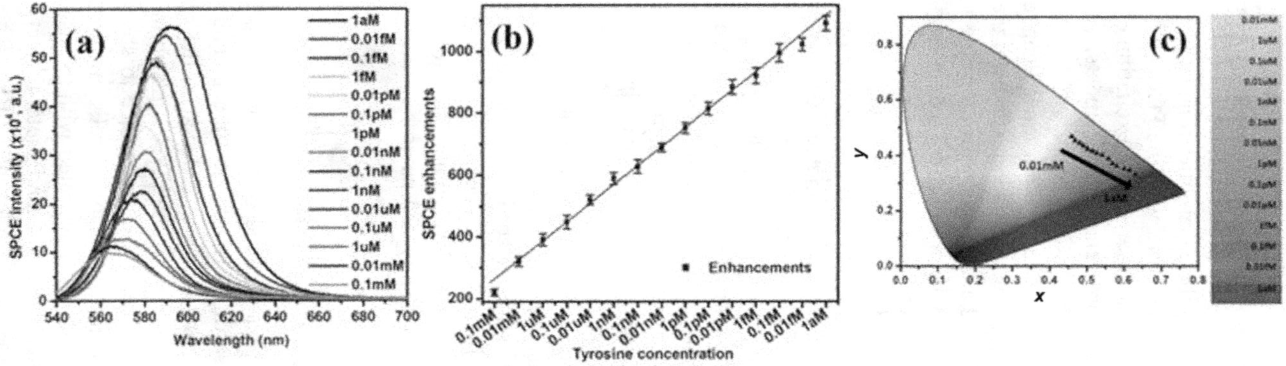

FIGURE 8.10 Sensing performed using Kollidon® based AgAu nanohybrids. (a) SPCE signal intensity for different concentrations of tyrosine from 0.1 mM to 1 aM. (b) SPCE enhancements for different concentrations of tyrosine from 0.1 mM to 1 aM. (c) Chromaticity diagram presenting a shift in color of the emission observed with different tyrosine concentrations. Reprinted with permission from Reference [29], Copyright 2022, Elsevier.

has been observed between the mobile-phone-based sensing and the conventional Ocean optics based sensors.

8.7 BIOSENSING USING LERDS IN THE PCCE PLATFORM

In this section, we will briefly discuss the capabilities rendered by the PCCE platform made up of 1DPhC. In our pioneering work with the PCCE platform, the dielectric thickness of the PVA layer as well the laser source excitation conditions have been optimized to achieve augmented fluorescence enhancements. While the SPCE platform yielded 10-fold enhancements, the PCCE platform yielded 40-fold enhancements in the fluorescence [46]. This also reflected upon cavity nanoengineering, where 200-fold PCCE enhancements were observed using AgNPs in the case of the PCCE platform vis-à-vis 60-fold SPCE enhancements with the same material [6,9]. This augmented PCCE enhancements has been utilized for femtomolar sensing of aluminum ions upon its chemical interaction with ARS fluorescent molecules [6].

Furthermore, novel nanogeometries have been explored in the SPCE and PCCE platform to further boost the photo-plasmonic coupling efficiency[47–49]. Ag nanoprisms of varying edge lengths yielded tunable PCCE enhancements and the optimized conditions for maximum coupling was determined. Such variants were considered for iodide ion sensing based on their close interfacial interaction with sharp-edged AgNPrsat nanodimensions [4]. The sharp surfaces of AgNPs interact with the iodide ions so as to make the surface blunt/more curved. Consequently, the plasmonic hotspot intensity decreases thereby resulting in dwindled PCCE enhancements. The proposed sensor could differentiate hypo- and hyperthyroidism based on the concentrations of the iodide ions used [4]. Moreover, the hybrid metal-dielectric nanoengineering carried out using Nd_2O_3 nanorods and Ag nanoprims yielded 1300-fold PCCE enhancements. This unprecedented enhancement was utilized for single molecular detection of cortisol molecule, which was chemically linked to rhodamine B moiety [4]. This strategy of indirect detection of cortisol molecule could be extrapolated towards development of effective ELISA strategies for sensing cortisol and other biologically and environmentally relevant molecules and ions.

8.8 CONCLUDING REMARKS

This chapter provides a detailed background of the nanoplasmonics arena and the related applications with the use of LERDs. The fundamental aspects of surface plasmons, metal-enhanced fluorescence are introduced to the readers followed by the fabrication and functioning of the SPCE and PCCR platform. Further, the novel nanoengineering strategies are discussed with relevance to spacer, cavity, and ext. cavity nanointerfaces. The nanostructure activity relationship towards achieving dequenched emission from LERDs is demonstrated with the use of gold nanostars. Furthermore, the utility of metal, dielectric, and metal-dielectric nanohybrids for biosensing application is discussed using relevant LERDs. The rational approach towards search for better plasmonic materials from metal NPs to pristine metal nanoassemblies to metal-dielectric nanoassemblies to lossless dielectric NPs with anisotropic geometries are captured with specific examples. The case studies are substantiated with appropriate sensing approaches achieved using abundant plasmonic hotspots. The importance of bimetallic or heterometallic nanohybrids to augment the performance of LERDs is further analyzed using frugal approaches wherein the utility of biopolymers, such as soluplus®, kollidon®, and gelucire® in photo-plasmonic biosensing frameworks are highlighted. The unprecedented enhancement in the performance of LERDs using the PCCE platform is explained with the help of iodide ion and cortisol molecule sensing at ultra-low concentrations. We firmly believe that the concepts discussed in this chapter along with the relevant case studies provided to support the claims made would be of immediate relevance to the broad audience of photonics, plasmonics, and biosensors.

ACKNOWLEDGEMENT

Authors acknowledge support from Tata Education and Development Trust [TEDT/MUM/HEA/SSSIHL/2017-2018/0069-RM-db], Prasanthi Trust, Inc., U.S. (22-06-2018), DST-Technology Development Program (IDP/MED/ 19/2016), Life Sciences Research Board (LSRB), and DST-Inspire Fellowship (IF180392), Govt. of India. We especially acknowledge SSSIHL-CRIF for extending the usage of the required instrumentation facility. Guidance from Bhagawan Sri Sathya Sai Baba is gratefully acknowledged.

REFERENCES

[1] Puebla-Hellmann, G., Venkatesan, K., Mayor, M., & Lörtscher, E. (2018). Metallic nanoparticle contacts for high-yield, ambient-stable molecular-monolayer devices. *Nature*, *559*(7713), 232–235.

[2] Singh, D., Rajput, D., & Kanvah, S. (2022). Fluorescent probes for targeting endoplasmic reticulum: design strategies and their applications. *Chemical Communications*, *58*(15), 2413–2429.

[3] Rai, B., Malmberg, R., Srinivasan, V., Ganesh, K.M., Kambhampati, N.S.V., Andar, A., Rao, G., Sanjeevi, C.B., Venkatesan, K. and Ramamurthy, S.S., 2021. Surface plasmon-coupled dual emission platform for ultrafast oxygen monitoring after SARS-CoV–2 infection. *ACS Sensors*, *6*(12), pp.4360–4368.

[4] Bhaskar, S., Singh, A.K., Das, P., Jana, P., Kanvah, S., Bhaktha B.N., S., & Ramamurthy, S.S. (2020). Superior resonant nanocavities engineering on the photonic crystal-coupled emission platform for the detection of femtomolar iodide and zeptomolar cortisol. *ACS Applied Materials &Interfaces*, *12*(30), 34323–34336.

[5] Rai, A., Bhaskar, S., Reddy, N., & Ramamurthy, S.S. (2021). Cellphone-aided attomolar zinc ion detection using silkworm protein-based nanointerface engineering in a plasmon-coupled dequenched emission platform. *ACS Sustainable Chemistry & Engineering*, *9*(44), 14959–14974.

[6] Bhaskar, S., Das, P., Srinivasan, V., Bhaktha B.N., S., & Ramamurthy, S.S. (2020). Bloch surface waves and internal optical modes-driven photonic crystal-coupled emission platform for femtomolar detection of aluminum ions. *The Journal of Physical Chemistry C*, *124*(13), 7341–7352.

[7] Bhaskar, S., Moronshing, M., Srinivasan, V., Badiya, P.K., Subramaniam, C., & Ramamurthy, S.S. (2020). Silver soret nanoparticles for femtomolar sensing of glutathione in a surface plasmon-coupled emission platform. *ACS Applied Nano Materials*, *3*(5), 4329–4341.

[8] Bhaskar, S., Das, P., Moronshing, M., Rai, A., Subramaniam, C., Bhaktha, S.B., & Ramamurthy, S.S. (2021). Photoplasmonic assembly of dielectric-metal, Nd2O3-Gold soret nanointerfaces for dequenching the luminophore emission. *Nanophotonics*, *10*(13), 3417–3431.

[9] Bhaskar, S., Das, P., Srinivasan, V., Bhaktha, S.B., & Ramamurthy, S.S. (2022). Plasmonic-silver sorets and dielectric-Nd_2O_3 nanorods for ultrasensitive photonic crystal-coupled emission. *Materials Research Bulletin*, *145*, 111558.

[10] Wang, D., Malmberg, R., Pernik, I., Prasad, S.K., Roemer, M., Venkatesan, K., Schmidt, T.W., Keaveney, S.T. & Messerle, B.A., 2020. Development of tethered dual catalysts: synergy between photo-and transition metal catalysts for enhanced catalysis. *Chemical Science*, *11*(24), pp.6256–6267.

[11] Jana, S., Mukherjee, S., Bhaktha B.N., S., & Ray, S.K. (2021). Plasmonic Silver Nanoparticle-Mediated Enhanced Broadband Photoresponse of Few-Layer Phosphorene/Si Vertical Heterojunctions. *ACS Applied Materials & Interfaces*, *14*(1), 1699–1709.

[12] Rai, A., Bhaskar, S., & Ramamurthy, S.S. (2021). Plasmon-coupled directional emission from soluplus-mediated AgAu nanoparticles for attomolar sensing using a smartphone. *ACS Applied Nano Materials*, *4*(6), 5940–5953.

[13] Lakowicz, J. R., Ray, K., Chowdhury, M., Szmacinski, H., Fu, Y., Zhang, J., & Nowaczyk, K. (2008). Plasmon-controlled fluorescence: a new paradigm in fluorescence spectroscopy. *Analyst*, *133*(10), 1308–1346.

[14] Rai, A., Bhaskar, S., Ganesh, K.M., & Ramamurthy, S.S. (2022). Gelucire®-mediated heterometallic AgAu nanohybrid engineering for femtomolar cysteine detection using smartphone-based plasmonics technology. *Materials Chemistry and Physics*, *279*, 125747.

[15] Gryczynski, I., Malicka, J., Gryczynski, Z., & Lakowicz, J.R. (2004). Radiative decay engineering 4. Experimental studies of surface plasmon-coupled directional emission. *Analytical Biochemistry*, *324*(2), 170–182.

[16] Rai, A., Bhaskar, S., Ganesh, K.M., & Ramamurthy, S.S. (2021). Engineering of coherent plasmon resonances from silver soret colloids, graphene oxide and Nd2O3 nanohybrid architectures studied in mobile phone-based surface plasmon-coupled emission platform. *Materials Letters*, *304*, 130632.

[17] Arathi, P.J., Bhaskar, S., & Ramanathan, V. (2018). The photocatalytic role of electrodeposited copper on pencil graphite. *Physical Chemistry Chemical Physics*, *20*(5), 3430–3432.

[18] Moronshing, M., Bhaskar, S., Mondal, S., Ramamurthy, S.S., &Subramaniam, C. (2019). Surface-enhanced Raman scattering platform operating over wide pH range with minimal chemical enhancement effects: Test case of tyrosine. *Journal of Raman Spectroscopy*, *50*(6), 826–836.

[19] Li, R., Deng, X., Liu, F., Yang, Y., Zhang, Y., Reddy, N., Liu, W., Qiu, Y. & Jiang, Q., 2020. Three-dimensional rope-like and cloud-like nanofibrous scaffolds facilitating in-depth cell infiltration developed using a highly conductive electrospinning system. *Nanoscale*, *12*(32), pp.16690–16696.

[20] Arathi, P.J., Bhaskar, S., & Ramanathan, V. (2018). Disulphide linkage: to get cleaved or not? Bulk and nano copper based SERS of cystine. *Spectrochimica Acta Part A: Molecular and Biomolecular Spectroscopy*, *196*, 229–232.

[21] Jana, S., Mukherjee, S., Ghorai, A., Bhaktha, S.B., & Ray, S.K. (2020). Negative thermal quenching and size-dependent optical characteristics of highly luminescent phosphorene nanocrystals. *Advanced Optical Materials*, *8*(12), 2000180.

[22] Schwarz, F., Kastlunger, G., Lissel, F., Riel, H., Venkatesan, K., Berke, H., Stadler, R. & Lörtscher, E., 2014. High-conductive organometallic molecular wires with delocalized electron systems strongly coupled to metal electrodes. *Nano Letters*, *14*(10), pp.5932–5940.

[23] Bhaskar, S., & Ramamurthy, S.S. (2019). Mobile phone-based picomolar detection of tannic acid on Nd_2O_3 nanorod–metal thin-film interfaces. *ACS Applied Nano Materials*, *2*(7), 4613–4625.

[24] Rathnakumar, S., Bhaskar, S., Rai, A., Saikumar, D.V., Kambhampati, N.S.V., Sivaramakrishnan, V., & Ramamurthy, S.S. (2021). Plasmon-coupled silver nanoparticles for mobile phone-based attomolar sensing of mercury ions. *ACS Applied Nano Materials*, *4*(8), 8066–8080.

[25] Bhaskar, S., Patra, R., Kowshik, N.C.S., Ganesh, K.M., Srinivasan, V., & Ramamurthy, S.S. (2020). Nanostructure effect on quenching and dequenching of quantum emitters on surface plasmon-coupled interface: A comparative analysis using gold nanospheres and nanostars. *Physica E: Low-dimensional Systems and Nanostructures*, *124*, 114276.

[26] Cao, S.H., Cai, W.P., Liu, Q., & Li, Y.Q. (2012). Surface plasmon–coupled emission: what can directional fluorescence bring to the analytical sciences? *Annual Review of Analytical Chemistry*, *5*, 317–336.

[27] Sadrolhosseini, A.R., Shafie, S., & Fen, Y.W. (2019). Nanoplasmonic sensor based on surface plasmon-coupled emission. *Applied Sciences*, *9*(7), 1497.

[28] Bhaskar, S., Kowshik, N.C.S., Chandran, S.P., & Ramamurthy, S.S. (2020). Femtomolar detection of spermidine using au decorated SiO2 nanohybrid on plasmon-coupled extended cavity nanointerface: a smartphone-based fluorescence dequenching approach. *Langmuir*, *36*(11), 2865–2876.

[29] Rai, A., Bhaskar, S., Ganesh, K.M., & Ramamurthy, S.S. (2022). Cellphone-based attomolar tyrosine sensing based on Kollidon-mediated bimetallic nanorod in plasmon-coupled directional and polarized emission architecture. *Materials Chemistry and Physics*, *285*, 126129.

[30] Dutta Choudhury, S., Badugu, R., Ray, K., & Lakowicz, J.R. (2015). Directional emission from metal–dielectric–metal structures: effect of mixed metal layers, dye location, and dielectric thickness. *The Journal of Physical Chemistry C*, *119*(6), 3302–3311.

[31] Pan, X.H., Cao, S.H., Chen, M., Zhai, Y.Y., Xu, Z.Q., Ren, B., & Li, Y.Q. (2020). In situ and sensitive monitoring of configuration-switching involved dynamic adsorption by surface plasmon-coupled directional enhanced Raman scattering. *Physical Chemistry Chemical Physics*, *22*(22), 12624–12629.

[32] Cao, S.H., Weng, Y.H., Xie, K.X., Wang, Z.C., Pan, X.H., Chen, M., Zhai, Y.Y., Xu, L.T. and Li, Y.Q., 2019. Surface plasmon coupled fluorescence-enhanced interfacial "molecular beacon" to probe biorecognition switching: an efficient, versatile, and facile signaling biochip. *ACS Applied Bio Materials*, *2*(2), pp. 625–629.

[33] Zhao, Y., Liu, Y.H., Cao, S.H., Ajmal, M., Zhai, Y.Y., Pan, X.H., Chen, M. and Li, Y.Q., 2020. Excitation–emission synchronization-mediated directional fluorescence: insight into plasmon-coupled emission at vibrational resolution. *The Journal of Physical Chemistry Letters*, *11*(7), pp. 2701–2707.

[34] Bhaskar, S., & Ramamurthy, S.S. (2021, April). High Refractive Index Dielectric TiO 2 and Graphene Oxide as Salient Spacers for >300-fold Enhancements. In *2021*

IEEE International Conference on Nanoelectronics, Nanophotonics, Nanomaterials, Nanobioscience & Nanotechnology (5NANO) (pp. 1–6). IEEE.

[35] Rai, A., Bhaskar, S., Mohan, G.K., & Ramamurthy, S.S. (2022). Biocompatible Gellucire® inspired bimetallic nanohybrids for augmented fluorescence emission based on graphene oxide interfacial plasmonic architectures. *ECS Transactions, 107*(1), 4527.

[36] Bhaskar, S., Rai, A., Mohan, G.K., & Ramamurthy, S.S. (2022). Mobile phone camera-based detection of surface plasmon-coupled fluorescence from streptavidin magnetic nanoparticles and graphene oxide hybrid nanointerface. *ECS Transactions, 107*(1), 3223.

[37] Bhaskar, S., Jha, P., Subramaniam, C., & Ramamurthy, S.S. (2021). Multifunctional hybrid soret nanoarchitectures for mobile phone-based picomolar Cu^{2+} ion sensing and dye degradation applications. *Physica E: Low-dimensional Systems and Nanostructures, 132*, 114764.

[38] Bhaskar, S., Visweswar Kambhampati, N.S., Ganesh, K.M., Srinivasan, V., & Ramamurthy, S.S. (2021). Metal-free, graphene oxide-based tunable soliton and plasmon engineering for biosensing applications. *ACS Applied Materials & Interfaces, 13*(14), 17046–17061.

[39] Chowdhury, M.H., Ray, K., Geddes, C.D., & Lakowicz, J.R. (2008). Use of silver nanoparticles to enhance surface plasmon-coupled emission (SPCE). *Chemical Physics Letters, 452*(1–3), 162–167.

[40] Jahani, S., & Jacob, Z. (2016). All-dielectric metamaterials. *Nature Nanotechnology, 11*(1), 23–36.

[41] Naik, G.V., Shalaev, V.M., & Boltasseva, A. (2013). Alternative plasmonic materials: beyond gold and silver. *Advanced Materials, 25*(24), 3264–3294.

[42] Moronshing, M., & Subramaniam, C. (2018). Room temperature, multiphasic detection of explosives, and volatile organic compounds using thermodiffusion driven soret colloids. *ACS Sustainable Chemistry & Engineering, 6*(7), 9470–9479.

[43] Mondal, S., & Subramaniam, C. (2020). Xenobiotic contamination of water by plastics and pesticides revealed through real-time, ultrasensitive, and reliable surface-enhanced Raman scattering. *ACS Sustainable Chemistry & Engineering, 8*(20), 7639–7648.

[44] Duan, H., Wang, D., & Li, Y. (2015). Green chemistry for nanoparticle synthesis. *Chemical Society Reviews, 44*(16), 5778–5792.

[45] Falsini, S., Bardi, U., Abou-Hassan, A., & Ristori, S. (2018). Sustainable strategies for large-scale nanotechnology manufacturing in the biomedical field. *Green Chemistry, 20*(17), 3897–3907.

[46] Bhaskar, S., Lis, S. S. M., Kanvah, S., Bhaktha, BN. S., &Ramamurthy, S.S. (2022). Single-molecule cholesterol sensing by integrating silver nanowire propagating plasmons and graphene oxide π-plasmons on a photonic crystal-coupled emission platform. *ACS Applied Optical Materials, 1*(1), 159–172.

[47] Bhaskar, S., Thacharakkal, D., Ramamurthy, S.S.,& Subramaniam, C. (2023). Metal–dielectric interfacial engineering with mesoporous nano-carbon florets for 1000-fold fluorescence enhancements: smartphone-enabled visual detection of perindopril erbumine at a single-molecular level. *ACS Sustainable Chemistry & Engineering, 11*(1), 78–91.

[48] Bhaskar, S., Srinivasan, V., & Ramamurthy, S.S. (2023). Nd_2O_3-Ag nanostructures for plasmonic biosensing, antimicrobial, and anticancer applications. *ACS Applied Nano Materials. 6*(2), 1129–1145.

[49] Rai, A., Bhaskar, S., Ganesh, K.M., & Ramamurthy, S.S. (2022). Hottest hotspots from the coldest cold: welcome to nano 4.0. *ACS Applied Nano Materials. 5*(9), 12245–12264.

Index

For Product Safety Concerns and Information please contact our EU
representative GPSR@taylorandfrancis.com
Taylor & Francis Verlag GmbH, Kaufingerstraße 24, 80331 München, Germany

www.ingramcontent.com/pod-product-compliance
Lightning Source LLC
Chambersburg PA
CBHW060443240326
41598CB00087B/3208